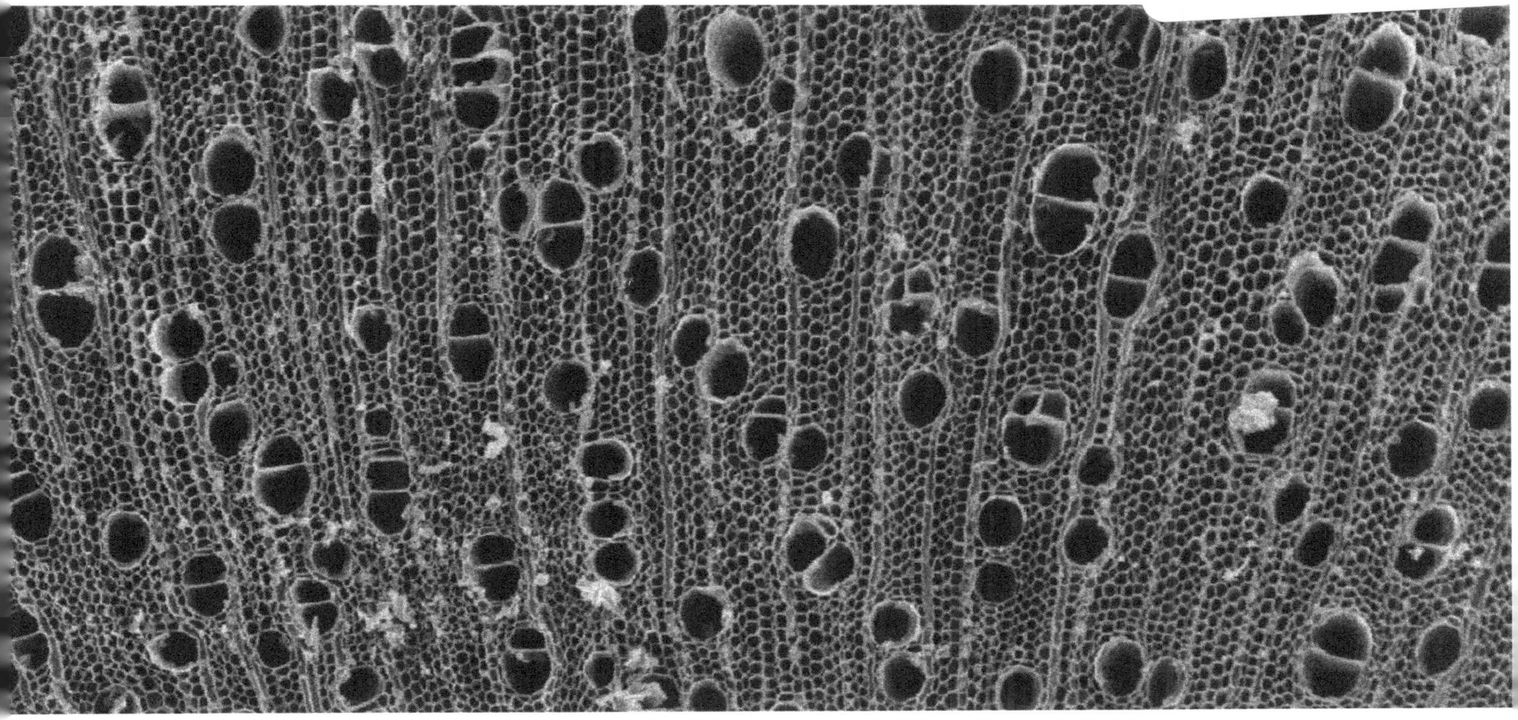

BAR INTERNATIONAL SERIES 3072 | 2022

Past Environments and Plant Use in Holocene Southern Africa

A study of charcoal and seed remains from the Late Stone Age sites of Toteng (Botswana), Leopard Cave and Geduld (Namibia)

Malebogo Mvimi

BAR
PUBLISHING

Published in 2022 by
BAR Publishing, Oxford

BAR International Series 3072

African Archaeology, volume 95
Past Environments and Plant Use in Holocene Southern Africa

ISBN 978 1 4073 5717 1 paperback
ISBN 978 1 4073 5718 8 e-format

DOI https://doi.org/10.30861/9781407357171

A catalogue record for this book is available from the British Library

COVER IMAGE *Transversal phase of* Commiphora glandulosa *charcoal*

BAR
PUBLISHING

BAR titles are available from:

BAR Publishing
122 Banbury Rd, Oxford, OX2 7BP, UK
EMAIL info@barpublishing.com
PHONE +44 (0)1865 310431
FAX +44 (0)1865 316916
www.barpublishing.com

Titles in the Series

No 1 **Niger Delta** *Aspects of its Prehistoric Economy and Culture* by Nwanna Nzewunwa, BAR S75, 1980,ISBN 9780860540830

No 2 **Prehistoric Investigations in the Region of Jenne Mali** *Study in the Development of Urbanism in the Sahel, Parts 1 and 2* by Susan Keech McIntosh and Roderick J. McIntosh, BAR S89, 1980, ISBN 9780860541035

No 3 **Off-Site Archaeology and Human Adaptation in Eastern Africa** *An Analysis of Regional Artefact Density inthe Amboseli, Southern Kenya* by Robert Foley, BAR S97, 1981, ISBN 9780860541141

No 4 **Later Pleistocene Cultural Adaptations in Sudanese Nubia** by Yousif Mukhtar el Amin, BAR S114, 1981, ISBN 9780860541349

No 5 **Settlement Patterns in the Iron Age of Zululand** *An Ecological Interpretation* by Martin Hall, BAR S119, 1981, ISBN 9780860541431

No 6 **The Neolithic Period in the Sudan, c.6000-2500 B.C.** by Abbas S.Mohammed-Ali, BAR S139, 1982,ISBN 9780860541707

No 7 **History and Ethnoarchaeology in Eastern Nigeria** *A Study of Igbo-Igala relations with special reference to the Anambra Valley* by PhilipAdigwe Oguagha andAlex Ikechukwu Okpoko, BAR S195, 1984, ISBN 9780860542490

No 8 **Meroitic Settlement in Central Sudan** *An Analysis of Sites in the Nile Valley and the Western Butana* by Khidir Abdelkarim Ahmed, BAR S197, 1984, ISBN 9780860542520

No 9 **Economy and Technology in the Late Stone Age of Southern Natal** by Charles Cable, BAR S201, 1984, ISBN 9780860542582

No 10 **Frontiers** *Southern African Archaeology Today* Edited by M. Hall, G. Avery, D.M. Avery, M.L. Wilson,A.J.B. Humphreys, BAR S207, 1984, ISBN 9780860542681

No 11 **Archaeology and History in Southern Nigeria** *The ancient linear earthworks of Benin and Ishan, Parts 1 and 2* by P.J. Darling, BAR S215, 1984, ISBN 9780860542759

No 12 **The Later Stone Age in Southernmost Africa** by Janette Deacon, BAR S213, 1984, ISBN 9780860542766

No 13 **Fisher-Hunters and Neolithic Pastoralists in East Turkana Kenya** by John Webster Barthelme, BAR S254, 1985, ISBN 9780860543251

No 14 **The Archaeology of Central Darfur (Sudan) in the 1st Millennium A.D.** by Ibrahim Musa Mohammed, BAR S285, 1986, ISBN 9780860543671

No 15 **Stable Carbon Isotopes and Prehistoric Diets in the South-Western Cape Province, South Africa** by Judith Sealy, BAR S293, 1986, ISBN 9780860543763

No 16 **L'art rupestre préhistorique des massifs centraux sahariens** by Alfred Muzzolini, BAR S318, 1986, ISBN 9780860544067

No 1 **Spheroids and Battered Stones in the African Early and Middle Stone Age** by Pamela R Willoughby, BAR S321, 1987, ISBN 9780860544104

No 18 **The Royal Crowns of Kush** *A study in Middle Nile Valley regalia and iconography in the 1st Millennia B.C. andA.D.* by Lázló Török, BAR S338, 1987, ISBN 9780860544326

No 19 **The Later Stone Age of the Drakensberg Range and its Foothills** by H. Opperman, BAR S339, 1987, ISBN 9780860544371

No 20 **Socio-Economic Differentiation in the Neolithic Sudan** by Randi Haaland, BAR S350, 1987, ISBN 9780860544531

No 21 **Later Stone Age Settlement Patterns in the Sandveld of the South-Western Cape Province, South Africa** by Anthony Manhire, BAR S351, 1987, ISBN 9780860544548

No 22 **L'art rupestre du Fezzan septentrional (Libye) Widyan Zreda et Tarut (Wadi esh-Shati)** by Jean-Loïc Le Quellec, BAR S365, 1987, ISBN 9780860544739

No 23 **Archaeology and Environment in the Libyan Sahara** *The excavations in the Tadrart Acacus, 1978-1983* Editedby Barbara E Barich, BAR S368, 1987, ISBN 9780860544746

No 24 **The Early Farmers of Transkei, Southern Africa before A.D. 1870** by J M Feely, BAR S378, 1987, ISBN 9780860544869

No 25 **Later Stone Age Hunters and Gatherers in the Southern Transvaal** *Social and ecological interpretation* byLyn Wadley, BAR S380, 1987, ISBN 9780860544920

No 26 **Prehistoric Cultures and Environments in the Late Quaternary of Africa** Edited by John Bower and David Lubell, BAR S405, 1988 ISBN 9780860545200

No 27 **Zooarchaeology in the Middle Nile Valley** *A Study of four Neolithic Sites near Khartoum* by Ali Tigani El Mahi, BAR S418, 1988, ISBN 9780860545392

No 28 **L'Ancienne Métallurgie du Fer à Madagascar** by Chantal Radimilahy, BAR S422, 1988, ISBN 9780860545446No 29 **El Geili** *The History of a Middle Nile Environment, 7000 B.C.-A.D. 1500* Edited by I. Caneva, BAR S424, 1988, ISBN 9780860545484

No 30 **The Ethnoarchaeology of the Zaghawa of Darfur (Sudan)** *Settlement and Transcience* by Natalie Tobert, BAR S445, 1988, ISBN 9780860545743

No 31 **Shellfish in Prehistoric Diet** *Elands Bay, S.W. Cape Coast, South Africa* by W. F. Buchanan, BAR S455, 1988, ISBN 9780860545842

No 32 **Houlouf I** *Archéologie des sociétés protohistoriques du Nord-Cameroun* by Augustin Holl, BAR S456, 1988, ISBN 9780860545866

No 33 **The Predynastic Lithic Industries of Upper Egypt** by Liane L. Holmes, BAR S469, 1989, ISBN 9780860546016 No 34 **Fishing Sites of North and East Africa in the Late Pleistocene and Holocene** *Environmental Change and Human Adaptation* by Kathlyn Moore Stewart, BAR S521, 1989, ISBN 9780860546627

No 35 **Plant Domestication in the Middle Nile Basin** *An Archaeoethnobotanical Case Study* by Anwar Abdel-Magid, BAR S523, 1989, ISBN 9780860546641

No 36 **Archaeology and Settlement in Upper Nubia in the 1st Millennium A.D.** by David N. Edwards, BAR S537, 1989, ISBN 9780860546825

No 37 **Prehistoric Settlement and Subsistence in the Kaduna Valley, Nigeria** by Kolawole David Aiyedun & Thurstan Shaw, BAR S541, 1989, ISBN 9780860546849

No 38 **The Archaeology of the Meroitic State** *New perspectives on its social and political organisation* by David N. Edwards, BAR S640, 1996, ISBN 9780860548256

No 39 **Islam, Archaeology and History** *Gao Region (Mali) ca. AD 900 - 1250* by Timothy Insoll, BAR S647, 1996, ISBN 9780860548324

No 40 **State Formation in Egypt** *Chronology and society* by Toby A. H. Wilkinson, BAR S651, 1996,ISBN 9780860548386

No 41 **Recherches archéologiques sur la capitale de l'empire de Ghana** *Etude d'un secteur d'habitat à Koumbi Saleh, Mauritanie. Campagnes II-III-IV-V (1975-1976)-(1980-1981)* by S. Berthier, BAR S680, 1997, ISBN 9780860548683

No 42 **The Lower Palaeolithic of the Maghreb** *Excavations and analyses at Ain Hanech, Algeria* by Mohamed Sahnouni, BAR S689, 1998, ISBN 9780860548751

No 43 **The Waterberg Plateau in the Northern Province, Republic of South Africa, in the Later Stone Age** by Maria M. Van der Ryst, BAR S715, 1998, ISBN 9780860548935

No 44 **Cultural Succession and Continuity in S.E. Nigeria** *Excavations in Afikpo* by V. Emenike Chikwendu, BAR S734, 1998, ISBN 9780860549215

No 45 **The Emergence of Food Production in Ethiopia** by Tertia Barnett, BAR S763, 1999, ISBN 9780860549710

No 46 **Sociétés préhistoriques et Mégalithes dans le Nord-Ouest de la République Centrafricaine** by Étienne Zangato, BAR S768, 1999, ISBN 9780860549802

No 47 **Ethnohistoric Archaeology of the Mukogodo in North-Central Kenya** *Hunter-gatherer subsistence and the transition to pastoralism in secondary settings* by Kennedy K. Mutundu, BAR S775, 1999, ISBN 9780860549901

No 48 Échanges **et contacts le long du Nil et de la Mer Rouge dans l'époque protohistorique (IIIe et IIe millénaires avant J.-C.)** *Une synthèse préliminaire* by Andrea Manzo, BAR S782, 1999, ISBN 9781841710020

No 49 **Ethno-Archaeology in Jenné, Mali** *Craft and status among smiths, potters and masons* by Adria LaViolette, BAR S838, 2000, ISBN 9781841710433

No 50 **Hunter-Gatherers and Farmers** *An enduring Frontier in the Caledon Valley, South Africa* by Carolyn R. Thorp, BAR S860, 2000, ISBN 9781841710617

No 51 **The Kintampo Complex** *The Late Holocene on the Gambaga Escarpment, Northern Ghana* by Joanna Casey, BAR S906, 2000, ISBN 9781841712024

No 52 **The Middle and Later Stone Ages in the Mukogodo Hills of Central Kenya** *A Comparative Analysis of Lithic Artefacts from Shurmai (GnJm1) and Kakwa Lelash (GnJm2) Rockshelters* by G-Young Gang, BAR S964, 2001, ISBN 9781841712512

No 53 **Darfur (Sudan) In the Age of Stone Architecture c. 1000 – 1750 AD** *Problems in historical reconstruction* by Andrew James McGregor, BAR S1006, 2001, ISBN 9781841712857

No 54 **Holocene Foragers, Fishers and Herders of Western Kenya** by Karega-Mũnene, BAR S1037, 2002, ISBN 9781841714172

No 55 **Archaeology and History in Ìlàrè District (Central Yorubaland, Nigeria) 1200-1900 A.D.** by Akinwumi O. Ogundiran, BAR S1090, 2002, ISBN 9781841714684

No 56 **Ethnoarchaeology in the Zinder Region, Republic of Niger** *The site of Kufan Kanawa* by Anne Haour, BAR S1133, 2003, ISBN 9781841715063

No 57 **Le Capsien typique et le Capsien supérieur** Évolution ou contemporanéité. Les données technologiques by Noura Rahmani, BAR S1187, 2003, ISBN 9781841715537

No 58 **Fortifications et urbanisation en Afrique orientale** by Stéphane Pradines, BAR S1216, 2004, ISBN 9781841715766

No 59 **Archaeology and Geoarchaeology of the Mukogodo Hills and Ewaso Ng'iro Plains, Central Kenya** by Frederic Pearl, BAR S1247, 2004, ISBN 9781841716077

No 60 **Islamic Archaeology in the Sudan** by Intisar Soghayroun Elzein, BAR S1289, 2004, ISBN 9781841716398

No 61 **An Ethnoarchaeological Study of Iron-Smelting Practices among the Pangwa and Fipa in Tanzania** by Randi Barndon, BAR S1308, 2004, ISBN 9781841716572

No 62 **Archaeology and History in North-Western Benin** by Lucas Pieter Petit, BAR S1398, 2005, ISBN 9781841718378
No 63 **Traditions céramiques, Identités et Peuplement en Sénégambie** *Ethnographie comparée et essai de reconstitution historique* by Moustapha Sall, BAR S1407, 2005, ISBN 9781841718507

No 64 **Changing Settlement Patterns in the Aksum-Yeha Region of Ethiopia: 700 BC – AD 850** by Joseph W. Michels, BAR S1446, 2005, ISBN 9781841718828

No 65 **Safeguarding Africa's Archaeological Past** *Selected papers from a workshop held at the School of Oriental and African Studies, University of London, 2001* Edited by Niall Finneran, BAR S1454, 2006, ISBN 9781841718927

No 66 **Excavations at Kasteelberg and the Origins of the Khoekhoen in the Western Cape, South Africa** by Andrew B. Smith, BAR S1537, 2006, ISBN 9781841719696

No 67 **Archéologie du Diamaré au Cameroun Septentrional** *Milieux et peuplements entre Mandara, Logone, Bénouéet Tchad pendant les deux derniers millénaires* by Alain Marliac, BAR S1549, 2006, ISBN 9781841719788

No 68 **Chasse et** élevage **dans la Corne de l'Afrique entre le Néolithique et les temps historiques** by Joséphine Lesur, BAR S1602, 2007, ISBN 9781407300191

No 69 **The Emergence of Social and Political Complexity in the Shashi-Limpopo Valley of Southern Africa, AD 900 to 1300** *Ethnicity, class and polity* by John Anthony Calabrese, BAR S1617, 2007, ISBN 9781407300290

No 70 **Archaeofaunal remains from the past 4000 years in Sahelian West Africa** *Domestic livestock, subsistence strategies and environmental changes* by Veerle Linseele, BAR S1658, 2007, ISBN 9781407300948

No 71 **Il Sahara centro-orientale Dalla Preistoria ai tempi dei nomadi Tubu / The Central-Oriental Sahara. From Prehistory to the times of the nomadic Tubus** by Vanni Beltrami, BAR S1667, 2007, ISBN 9781407301020

No 72 **Memory and the Mountain** *Environmental Relations of the Wachagga of Kilimanjaro and Implications for Landscape Archaeology* by Timothy A. R. Clack, BAR S1679, 2007, ISBN 9781407301174

No 73 **Archaeological Investigations of Iron Age Sites in the Mema Region, Mali (West Africa)** by Tereba Togola, BAR S1736, 2008, ISBN 9781407301785

No 74 **CurrentArchaeological Research in Ghana** Edited by Timothy Insoll, BAR S1847, 2008, ISBN 9781407303345No 75 **Holocene Prehistory of the Southern Cape, South Africa** *Excavations at Blombos Cave and the Blombosfontein Nature Reserve* by Christopher Stuart Henshilwood, BAR S1860, 2008, ISBN 9781407303437

No 76 **The Archaeology of Tanzanian Coastal Landscapes in the 6th to 15th Centuries AD** *The Middle Iron Age ofthe Region* by Edward John David Pollard, BAR S1873, 2008, ISBN 9781407303536

No 77 **Using Stone Tools: the Evidence from Aksum, Ethiopia** by Laurel Phillipson with an introduction by Rodolfo Fattovich, BAR S1926, 2009, ISBN 9781407304083

No 78 **An Archaeological and Geomorphological Survey of the Luangwa Valley, Zambia** by Dan Colton, BAR S2022, 2009, ISBN 9781407305974

No 79 **Historical Archaeologies of Nineteenth-Century Colonial Tanzania** *A comparative study* by Daniel Rhodes, BAR S2075, 2010, ISBN 9781407306360

No 80 **Material Knowledges, Thermodynamic Spaces and the Moloko Sequence of the Late Iron Age (AD 1300-1840) in Southern Africa** by Per Ditlef Fredriksen, BAR S2387, 2012, ISBN 9781407309798

No 81 **Métallurgie du fer et Sociétés africaines** *Bilans et nouveaux paradigmes dans la recherche anthropologique et archéologique* Edited by Caroline Robion-Brunner and Bruno Martinelli, BAR S2395, 2012, ISBN 9781407309880

No 82 **Butua and the End of an Era** *The effect of the collapse of the Kalanga state on ordinary citizens. An analysis of behaviour under stress* by Catharina van Waarden, BAR S2420, 2012, ISBN 9781407310190

No 83 **The Butana Group Ceramics and their Place in the Neolithic and Post-Neolithic of Northeast Africa** by Frank Winchell, BAR S2459, 2013, ISBN 9781407310671

No 84 **The Archaeology of the West Coast of South Africa** Edited by Antonieta Jerardino, Antonia Malan and David Braun, BAR S2526, 2013, ISBN 9781407311449

No 85 **The Ancient Red Sea Port of Adulis and the Eritrean Coastal Region** *Previous investigations and museum collections* by Chiara Zazzaro, BAR S2569, 2013, ISBN 9781407311906

No 86 **Metals from K2 and Mapungubwe, Middle Limpopo Valley** *A technological study of early second millennium material culture, with an emphasis on conservation* by Farahnaz Koleini, BAR S2653, 2014, ISBN 9781407312958

No 87 **Archaeology of Mound-Clusters in WestAfrica** byAugustin F. C. Holl, BAR S2660, 2014, ISBN 9781407313023No 88 **Paleoethnobotanical Study of Ancient Food Crops and the Environmental Context in North-East Africa, 6000 BC–AD 200/300** by Alemseged Beldados, BAR S2706, 2015, ISBN 9781407313573

No 89 **Population and Ceramic Traditions: Revisiting the Tana Ware of Coastal Kenya (7th-14th Century AD)** by Freda Nkirote M'Mbogori, BAR S2717, 2015, ISBN 9781407313702

No 90 **Swahili Urbanisation, Trade and Food Production** *Botanical perspectives from Pemba Island, Tanzania, AD 600-1500* by Sarah Walshaw, BAR S2755, 2015, ISBN 9781407314167

No 91 **The So Pots of Central Africa** *Memories of the past* by Graham Connah, 2019 ISBN 9781407316888

No 92 **Modelling Water Use at Great Zimbabwe** *An ethnohistoric, ethnoarchaeological, and GIS landscape analysisat an ancient African city* by Tendai Treddah Musindo, 2019, ISBN 9781407353975

No 93 **Exploitation and Management of Animal Resources During the Middle Iron Age** *Northern Limpopo Province, South Africa* by Claudia Abatino, 2020, ISBN 9781407357263

No 94 **Imprints of the Archaeology of Northern Nigeria** *Landscape, society and crafts around Kirfi, Bauchi region* by Abubakar Sule Sani, 2021, ISBN 9781407358925

Of Related Interest

Human-Environment Dynamics in the Aeolian Islands during the Bronze Age
A paleodemographic model
Claudia Speciale

Oxford, BAR Publishing, 2021 — BAR International Series **3052**

Forest Fires and Abandonment Patterns from the Aguada Culture, Northwest Argentina
A Paleoenvironmental Study Based on the Analysis of Microcharcoals in the Ambato Valley, Catamarca Province
Henrik B. Lindskoug

Oxford, BAR Publishing, 2016 — BAR International Series **2803**

Charcoals From the Past: Cultural and Palaeoenvironmental Implications
Proceedings of the Third International Meeting of Anthracology, Cavallino - Lecce (Italy), June 28th - July 1st 2004
Edited by Girolamo Fiorentino and Donatella Magri

Oxford, BAR Publishing, 2008 — BAR International Series **1807**

Charcoal Analysis: New Analytical Tools and Methods for Archaeology
Papers from the Table-Ronde held in Basel 2004
Edited by Alexa Dufraisse

Oxford, BAR Publishing, 2006 — BAR International Series **1483**

For more information, or to purchase these titles, please visit **www.barpublishing.com**

To Basathu Manta Tabushwa, thank you for taking me to school and for instilling in me a curiosity for plants.

Acknowledgements

The research from which this book emanated would not have been possible without the immense support from several bodies and individuals, who I would like to profoundly acknowledge. The book is an outcome of my PhD thesis; hence, naturally, I would like to express my earnest gratitude to my supervisors Drs. Margareta Tengberg, David Pleurdeau and Ethel Alluè for their colossal propositions, guidance and their inordinate confidence in my work. The same goes to my examiners, Professor Pikirayi and Professor Oosterbeek for their expedient comments and recommendations.

Mobility to and from the study sites, dating and material analysis were feasible thanks to my being a beneficiary of the International Doctorate in Quaternary and Prehistory, Erasmus Mundus Scholarship. I am grateful to the Ministry of Youth Empowerment, Sport & Culture Development in Botswana for issuing the necessary research permit for undertaking studies at the site of Toteng. My appreciation equally goes to the National Museums of Botswana and Namibia together with the National Research Council of Namibia, for availing the substantial archaeological material, besides all indispensable authorisations for data collection and loaning. Whereas the archaeobotany laboratories of the Musèum National d'histoire Naturelle, (MNHN) in Paris and the Catalan Institute of Human Paleoecology and Social Evolution (IPHES), Tarragona are where all the technical activities, data analysis, identification and post- production ensued. Thanks, in particular to Mr Michel Lemoine (MNHN) for charring the wood specimens well on time and for his guidance regarding photography. Likewise, I am thankful to the staff at the University of Frankfurt Archaeobotany Unit, for assisting with the identification of the charcoal material. I would like to give recognition to the University of Botswana History department; Through my affiliation with them, I enjoyed ample working space, fruitful conversations with colleagues and abundant supplementary departmental resources that facilitated the compilation of this book. To the Leopard Cave and Toteng teams as well as the Omandumba and Tubusis residents, thank you for making fieldwork and data collection pleasurable. Special mention goes to Maria Mwatondange for assisting with ethnobotanical data collection and translations of questions from English to Afrikaans. I am grateful to Prof Fred Morton for proofreading this script and for his constructive and insightful suggestions. On a more intimate note, I am glad to be surrounded by incredibly supportive friends, acquantances and colleagues. While unfortunately I cannot mention them one by one, they have in their respective ways believed in me, inspired me to keep going, and are forever my source of moral and emotional support. On that note, I am forever indebted to Gape, Mma Thebe and their particular families as well Miassa, David and *les petits frères* for according me the ambiance (counting lots of laughter), encouragement and abode, conducive enough for writing. Similarly, to my brother Bodu and MmaTjose for always believing in my potential. And finally, I incalculably thank Dr. Stefania Merlo, for her profuse intuitive contributions, believing in me, constantly checking on the progress of the book, encouragements and numerous selfless sacrifices she has made for me.

N.B: Appropriately, copyright permission was sought to reproduce material belonging to 3rd parties. However, while authorisation was granted for the larger majority, at the time of going to press, a few of them were not yet cleared despite numerous requests.

Contents

List of Figures xv
List of Tables xix
Abstract xxi

Introduction 1

1. Environmental Background of Southern Africa 5
1.1. Delimination of the study area 5
1.1.1. Geomorphology and soils 5
1.2. Past and present climatic conditions 8
1.3. Past and present vegetation 10
1.4. Fauna 12

2. The Late Stone Age with a Special Focus on the Studied Sites 15
2.1. The Late Pleistocene and the onset of the LSA in Southern Africa 15
2.1.1. Technical changes during the LSA 16
2.2. The Holocene LSA and its place among the history of the peopling of Southern Africa 19
2.3. Chronology and the place of herders: the advent of a Southern African 'Neolithic'? 19
2.3.1. Subsistence strategies 22
2.4. Leopard Cave (Erongo, Namibia) 22
2.4.1. Stratigraphy and dating 23
2.4.2. Lithic materials 27
2.4.3. Faunal remains 27
2.4.4. Human remains 29
2.4.5. Pigments and rock art 30
2.4.6. Present day environment 32
2.5. Toteng 1 (Botswana) 32
2.5.1. Stratigraphy and dating 33
2.5.2. Archaeological remains 33
2.5.3. Pottery 36
2.5.4. Lithics 36
2.5.5. Non lithic products 37
2.5.6. Faunal remains 37
2.5.7. Lake Ngami as an important water source 39
2.5.8. Present day landscape and vegetation 40
2.6. Geduld (Kunene, Namibia) 41
2.6.1. Stratigraphy and datings 41
2.6.2. Material cultures 41
2.6.3. Faunal remains 43
2.6.4. Botanical remains 44

3. Archaeobotanical Material and Methods 45
3.1. Description of the study material 45
3.2. Sampling methods 46
3.3. Leopard Cave 46
3.4. Geduld 48
3.5. Toteng 1 48
3.6. Laboratory methods and botanical identifications 49
3.7. Charcoal analysis 50
3.8. Seed analysis 50
3.9. Constitution of a reference collection 51

4. Plant Use Among Present-day Damara and San Populations in the Erongo Region, Namibia 55
4.1. Objectives of the ethnobotanical study 55
4.2. Description of the Damara village of Tubusis 55
4.2.1. The Damara ethnic group 56

4.2.2. Methodology ... 57
4.2.3. Description of plant uses for each household ... 58
4.2.4. General observations ... 60
4.3. The San Living Museum in the Omandumba farm ... 60
4.3.1. The San populations of Southern Africa ... 63
4.3.2. Plant use in the San community from Tsumkwe ... 64

5. Results ... 69
5.1. Description of identified taxa ... 69
5.2. Leopard Cave: results of the anthracological study ... 112
5.2.1. Results of the seed analysis ... 113
5.2.2. First observations on the macrobotanical record at Leopard Cave ... 113
5.3. Geduld: results of the anthracological study ... 115
5.3.1. Results of the seed analysis ... 117
5.3.2. First observations on the macrobotanical record at Geduld ... 119
5.4. Toteng: results of the anthracological study ... 119
5.5. Preliminary conclusions ... 120

6. Discussion ... 121
6.1. Leopard Cave's arboreal species from charcoal ... 121
6.2. Arboreal species from seed analysis ... 121
6.3. Geduld's arboreal species from charcoal ... 123
6.4. Arboreal species from seed analysis ... 123
6.5. Annual plants from seeds analysis ... 123
6.6. Toteng's arboreal species from charcoal ... 123
6.7. Making inferences from today's vegetation ... 124
6.8. Wild fauna to complement the environmental setting ... 125
6.9. Subsistence strategies and use of biological resources ... 126
6.10. The choice of species ... 126
6.11. Two exceptional species of Colophospermum mopane and Sclerocarya birrea ... 126
6.12. Seasonality of fruit collection and possible processing ... 127

Appendix 1: Wood Anatomy for Identification Purposes ... 131
Appendix 2: List of Identified Families Featuring in the Wood Atlas (According to Coates and Palgrave 2002; van Wyk and van Wyk 2007) ... 135
References ... 185

List of Figures

Figure 1. Map of Southern Africa with the localisation of the study sites 2

Figure 2. Map of Southern Africa with the localisation of the study sites and geographical features mentioned in the text 6

Figure 3. Major soil types of Southern Africa 7

Figure 4. Current annual precipitation over southern Africa 9

Figure 5. Biomes of Southern Africa 11

Figure 6. Map showing the density of main LSA sites in Southern Africa 16

Figure 7. Synthetic cultural chronology of Late MSA and LSA traditions in Southern Africa 17

Figure 8. Variances between Oakhurst and Wilton artefacts from southern Africa. 18

Figure 9. Later Stone Age sites of Southern Africa with early evidence of caprines 20

Figure 10. One of the proposed migration models showing routes from the west into Southern Africa possibly with livestock 21

Figure 11. Location of the site of Leopard Cave 23

Figure 12. View of Leopard cave in 2012 (l) and location of the excavations areas (r) 24

Figure 13. East-West transversal section and an example of recovered material from Leopard cave. 25

Figure 14. Stratigraphic section of 8/9 transversal bands (Squares NOP8/NOP9) 26

Figure 15. Radiocarbon ages of the upper Complex of Leopard Cave according to the elevation. (2 sigma, SHcal13) 28

Figure 16. Examples of lithic materials from Upper complex of Leopard Cave 29

Figure 17. Small bovid tooth from Upper Complex I of Leopard Cave, initially identified as caprine 30

Figure 18. Human zygomatic from the Upper Complex I 30

Figure 19. Giraffe figure as depicted in the rock art 30

Figure 20. A) In situ pestle (P) and grinding stone (G.S) associated with ferruginous colouring material. B) Pestle covered in red residue and ferruginous colouring material showing evidence of use-wear (C) 31

Figure 21. Current environment in close proximity to the site of Leopard Cave 31

Figure 22. Location of the Toteng sites, in relation to Lake Ngami (☆) with a view of the Toteng 1 sampling area in 2016 32

Figure 23. Location of the excavation areas (l) and view of the squares G-L excavation in 2003(r) 34

Figure 24. Stratigraphic sections from the six excavated squares and ages obtained for the Toteng 1 archaeological site 35

Figure 25. Bowl, possible bowl and possible jars of Bambata pottery from Toteng presenting comb-stamping, incisions and with red ochre 37

Figure 26. Lithics artifacts from Toteng 1, including backed drill and tools, as segments, trapezoid and distally backed points 38

Figure 27. Comparison between buffalo and cattle carpals from Toteng 38

Figure 28. Present day Lake Ngami 39

Figure 29. The current surroundings of Toteng, bottom left is part of Lake Ngami 40

Figure 30. Location of Geduld (Namibia) 41

Figure 31. Stratigraphic sequence of the Geduld excavation 42

Figure 32. Fruiting seasons of plants identified from Geduld 44

Figure 33. A mixture of carbonised and desiccated fruit remains from Geduld 45

Figure 34. General overview of the excavation area at Leopard Cave, red squares indicate zones that were sampled for archaeobotanical remains 46

Figure 35. W-E profile of square O8 in Leopard Cave 47

Figure 36. Excavation of Toteng 1 in 2016 49

Figure 37. Drawing of the stratigraphy revealed in 2016 (r) and section from the 2002–03 excavation (l), ... 49

Figure 38. Use of a reflected-light microscope for charcoal analysis in the archaeobotanical laboratory of the National Museum for Natural History, Paris 50

Figure 39. Areas in Namibia and Botswana where the reference material was collected 51

Figure 40. Labelled dried wood cuttings before charring 52

Figure 41. An example of a herbarium sheet with leaves and fruits/seeds from Vachellia erubescens 52

Figure 42. Location of Tubusis village in the Erongo region 56

Figure 43. View of part of Tubusis village and it's surroundings 56

Figure 44. Landscape opposite Tubusis village with granite foothills in the background and open Vachellia erubescens woodland with a grazing herd of goats 57

Figure 45. Map of Namibia showing the current distribution of Damara populations; orange colour on the left and blue"DAM" on the right 57

Figure 46. A Damara woman and man living in Tubusis 58

Figure 47. Mrs Meda !Kharuxab demonstrates the use of *Commiphora wildii* while being recorded on a voice recorder 59

Figure 48. Edible *Sporobolus sp.*, grass seeds that are collected from around ants' nests 59

Figure 49. Fencing around the yard of Mrs Esther Ganuses. Note the mixture of metal sheets and tree branches 60

Figure 50. Wood used for constructions and fuel stored beside the kitchen house in household 5 60

Figure 51. Exchanging with the informants at the San living Museum 63

Figure 52. The landscape in Omandumba West adjacent the San Living Museum 63

Figure 53. The distribution of San populations in Southern Africa ... 63

Figure 54. The location of Tsumkwe settlement in Namibia ... 65

Figure 55. Fire stick made from the Mongongo tree (*Schinziophyton rautanenii*) ... 67

Figure 56. A San man working *Sansevieria pearsonii* leaves into a string for bird trapping ... 67

Figure 57. Examples of archaeological seeds recovered from Geduld ... 112

Figure 58. Examples of archaeological seeds recovered from Geduld and Leopard Cave ... 112

Figure 59. Charcoal diagrams from Leopard Cave showing relative proportions of the identified taxa in Phases 1 & 2 ... 115

Figure 60. Charcoal diagram from Geduld showing relative proportions of the identified taxa ... 117

Figure 61. Charcoal diagram from Toteng showing relative proportions of the identified taxa ... 119

Figure 62. Mixed bushveld (Southeastern Botswana) ... 122

Figure 63. Thorny thicket corresponding to the Sweet Arid bushveld (Southeastern Botswana) ... 122

Figure 64. Vachellia bushveld near the site of Toteng ... 124

Figure 65. The current distribution of some of the major plant communities found on the sites ... 125

Figure 66. Vegetation of the Erongo showing Kudu (l) and zebra (c) in the thorny shrubland near Leopard Cave and an ostrich family in northwestern Botswana ... 126

Figure 67. An example of *Ximenia* for oil extraction in Angola and a fresh *Ximenia* fruit ... 129

Figure 68. Split *Ximenia* sp. endocarps ... 129

Figure A1. Cross-section of a tree showing the anatomical features ... 131

Figure A2. Wood phases and wood charcoal showing the below discussed features ... 132

Figure A3. A description of the thorn and leaf morphology for tree identification and differentiation ... 133

List of Tables

Table 1. Summary of general published charcoal studies in Africa from both the Pleistocene and Holocene periods 3

Table 2. Summary of the current geomorphological literature bearing on Southern Africa 6

Table 3. Physiognomy and main taxa of Southern African biomes 12

Table 4. Faunal taxa currently found in Southern Africa 13

Table 5. Summary of current chronological frame of MSA and LSA in Southern Africa 15

Table 6. Radiocarbon ages of the archaeological sequence of Leopard Cave 27

Table 7. OSL ages from Toteng 1 sequence done in 2016 36

Table 8. AMS ages from Toteng 1. Calibrated using OxCal 4.3 and IntCal 13 curve 36

Table 9. Toteng 1 artifact distribution within soil units from squares G-L 36

Table 10. Radiocarbon ages obtained at Geduld 42

Table 11. Faunal spectra from Geduld 43

Table 12. Categories of plant remains studied from each of the three sites 45

Table 13. List of samples studied from Geduld 48

Table 14. List of species in the reference collection with their family, Latin and vernacular name as well as the part of the plant collected (F/INFL – Flower/inflorescence; FR/S – Fruit/seed) 52

Table 15. Summary table of the uses of woody plants in the Tubusis Damara community 61

Table 16. Use of woody plants by the San community in Omandumba 66

Table 17. Comparison of the wood species used by the Damara and San communities 67

Table 18. Comparison of the anatomic characteristics of the archaeological charcoal 70

Table 19. Absolute numbers and frequencies of the different wood taxa identified from the occupation Phase I at Leopard Cave organised according to their layers and in alphabetical order 113

Table 20. Absolute numbers and frequencies of the different wood taxa identified from the occupation Phase IIat Leopard Cave organised according to the depth and in alphabetical order 114

Table 21. Result of the seed/fruit analysis at Leopard Cave 116

Table 22. Absolute numbers and frequencies of the different taxa identified at Geduld organised according to their square and level 117

Table 23. Result of the seed/fruit analysis at Geduld 118

Table 24. Absolute numbers and frequencies of the different taxa identified at Toteng organised according to their square and level 119

Table 25. Seasonality of fruit trees reflecting the macrobotanical spectra 128

Abstract

Through macrobotanical (wood, charcoal and seeds/fruit) analysis from three major Later Stone Age (LSA) sites located in the Kgalagadi Basin in southern Africa, this study is interested in reconstructing the environmental conditions during the Holocene in the region, in particular in the last two to three millennia. Initially this work couples archaeological macrobotanical analysis with the construction of modern reference material in an effort to trace the environmental/plant evolution as well as to comprehend socio-ecological and socio-environmental dynamics in Southern Africa during the late Holocene. The scope of this study covers the period spanning the last 3000–2000 years, with the main objective of understanding what relationships humans had with their environment at a time linked with the arrival or the appearance of the first herding practices in that part of Africa. These herding practices are believed to be accompanied by significant human movement from eastern or central Africa southwards. Favourable environmental conditions may have influenced their routes as well as settlement choices, and these are aspects that this archaeobotanical study aims to address. This study also employed an ethnographic approach, working with local communities in the Erongo region of Namibia, so as to make inferences to past vegetation utilisation practices while at the same time discerning and reconstituting past human activities.

Keywords: macrobotany, archaeobotany, wood charcoal analysis, seed analysis, Holocene, Southern Africa, Later Stone Age, ethnography, environmental reconstruction

A travers l'analyse de matériel macrobotanique (charbon et graines/fruits) issus de trois sites majeurs du *Later Stone Age* (LSA) d'Afrique australe, cette étude s'intéresse à reconstituer les conditions environnementales durant l'Holocène dans la région, en particulier au cours des deux à trois derniers millénaires. L'étude de ce matériel, accompagné de la constitution d'une collection moderne de références de la végétation du bassin Kgalagadi, vise à mieux cerner l'impact que l'environnement a pu avoir sur les sociétés humaines (subsistance, peuplements, etc.) à une période-clé du *(LSA)*, qui marque l'émergence des pratiques pastorales dans la région. L'étude des environnements est ici primordiale pour interpréter les conditions qui ont pu favoriser l'arrivée de ces premiers éleveurs et leurs lieux d'installation. En complément, une approche ethnographique a été menée, en travaillant avec les communautés locales dans la région de l'Erongo en Namibie, qui peuvent aider à comprendre et reconstituer les pratiques passées d'utilisation de la végétation.

Mots –clés: macrobotanique, archèobotanique, anthracologie, carpologie, Holocène, Afrique austral, Later Stone Age, ethnographie, reconstruction environnementale

Introduction

Southern Africa[1] is an important part of the scene in the narrative of human evolutionary history. The subcontinent has produced several early hominin taxa, from the Australopithecines and the early representatives of the genus Homo (Pickering et al., 2011) to the first anatomically Modern Humans. Today the region is home to several ethnicities among them the autochthonous Khoi San speaking groups whose members are considered to be the descendants of the first people of Southern Africa. Upon their early interactions with other groups, notably the proto-Khoekhoe, until recent times, their territory remained mainly restricted to the dry Kgalagadi basin, with a population essentially shared between modern day Namibia, Botswana, South Africa and Zimbabwe. Their ancestors existed in the region for thousands of years before the arrival of proto-Khoekhoe herders around 2000 years ago and then the subsequent arrival of Bantu-speaking groups from East and Central Africa around 1500 years ago (Pleurdeau et al. 2012, Mitchell 2002) followed by European colonisers in the 17th century. The lifestyle and particular adaptations of one of the last remaining hunter and gatherer populations of Southern Africa, evolving in an environment that is strongly marked by aridity, has attracted the attention of scholars and travellers since the 18th century. The San communities, even though they have now largely transitioned to become sedentary herders (Lee 1978) are often cited as an example in anthropological research (; Demangeon 2018, Penn 2013 Thondhlana and Shackleton 2015; Pamo 2011; Schuster et al. 2010). In contrast, our knowledge of the interactions between the ancestors of the San and their environment is still limited due to a lack of systematic studies of well-preserved remains from prehistoric sites.

This work aims at investigating the relationship between the Late Stone Age populations in Southern Africa and one element of this environment: the plant world. The study is based on the study of archaeobotanical macroremains (charcoal, wood and charred, desiccated and mineralised seed/fruit remains) from three sites: Leopard Cave and Geduld in present-day Namibia and Toteng in Botswana (Figure 1). Together they span a chronological period of more or less 2900 years, from ca. 3700 cal BP to ca. 800 cal BP.

The three sites are situated within the vast Kgalagadi basin covering a territory of approximately 2.5 million km^2 in Southern Africa and hosting a large diversity of landforms including rocky outcrops, extensive sand flats and major paleolake basins. New and advanced research methods have in the past few decades brought to light the diversity of the Southern African Quaternary paleoclimates and environments, predominantly by using multi-proxy approaches (Eitel et al. 2006). Close links exist between climate, topography, soil types, vegetation, flooding patterns and the temporal and spatial evolution of the environment. Major environmental changes, are widely viewed as a significant driver of human evolution in Southern Africa and in Africa as a whole[2] (Scott and Woodborne 2007; Chase & Meadows, 2007). One source of information on past plant formations and their evolution through time is provided by pollen studies. The survival of pollen is dependent however on the nature of deposits, and the dry environments that prevail in Southern Africa are usually unfavourable to the preservation of this type of organic markers (Wadley 2006). Despite the rarity of adequate "preservation containers" (Scott & Neumann, 2018), several studies have been conducted in the region since the 1950s allowing for reconstructions of the Pleistocene and Holocene vegetation cover in different areas. Scott el al. (2012) summarised these results allowing a closer view of pollen sequences over a wide region and providing clues to a better comprehension of paleoclimatic conditions. At the same time, the study has revealed the heterogeneous and fragmentary nature of the available data suggesting that supplementary, high-resolution pollen information in association with other palaeoenvironmental proxies is still much needed. The scarcity of sites, non-standardised methods of interpretation and low taxonomical resolution are among the factors that make it difficult to come up with a consistent and continuous regional palaeoenvironmental reconstruction. In order to compensate for bad preservation, alternative pollen sources notably from marine cores and hyrax[3] dung are also considered (for example Gil-Romera et al. 2006, 2007, Scott and Woodborne 2007b, Lim et al. 2016, Scott, Marais and Brook 2004). While these can be very rich in information on the past vegetation, their depositional history, and thus geographical and floral representativeness, is quite different from that of terrestrial and fluvial/lacustrine deposits.

[1] The southernmost region of the African continent, comprising the countries of Botswana, Angola, Lesotho, Malawi, Mozambique, Namibia, South Africa, Eswatini, Zambia and Zimbabwe is referred to in this work as Southern Africa or as the African subcontinent.

[2] During the Quaternary period, people have become dominant environmental agents. This development was influenced strongly by significant environmental changes such as seasonality of rainfall. In fact, a significant degree of precipitation reduction leading to extreme aridity prevailed in large areas of southern Africa, therefore having a major influence on landscape developments.

[3] Hyraxes are small herbivorous mammals belonging to the order of *Hyracoidea*. Four of these live in Africa, a fifth can also be found in the Middle East.

Figure 1. Map of Southern Africa with the localisation of the study sites. Map author: David Pleurdeau.

Another basis of information on past plant formations, especially those that grew near settlement sites and with which past populations interacted, is provided by the study of charred wood remains from archaeological contexts. The archaeobotanical field dealing with this material – anthracology (from the Greek word *anthrax* – charcoal) – is developed here, together with the study of seed and fruit remains.

The analysis of charred wood from archaeological settlements, camp fires and graves brings to light the human use of fuel and timber wood, thus addressing different aspects of the wood economy of past populations (Ludemann and Nelle 2017). Moreover, when certain methodological procedures are respected, charcoal remains stemming from domestic fires (used for cooking, heating, etc.) are likely to provide a representative reconstruction of the ligneous vegetation that was exploited for fuel over a longer period of time and thus indirectly of the local landscape dynamics (Ludemann and Nelle 2017). In some instances, charcoal can also be used as a proxy for reconstructing past climates (Thornton-Barnett 2013). Indeed the expression "timber does not fly" (Leroi-Gourhan in Neumann 1992) applies here as wood charcoal is usually found *in situ* and provides an idea of the environment within which it is found (Dotte-Sarout et al. 2015).

Charcoal studies are still rare in Southern Africa. One major reason for this is the lack of covered and/or stratified deposits, notably in rock shelters and caves, where organic remains are better preserved than they are in shallow open-air sites (Eichhorn and Jürgens 2003). The highest number of studies are concentrated in South Africa, Lesotho and Eswatini, a region where rock shelters are abundant (Esterhuysen 1996, Esterhuysen and Mitchell 1996, Thornton-Barnett 2013, Allsop 1998, Sievers 2006, 2016, Wadley et al. 2011, Shackleton and Prins 1992). Few anthracological studies have so far been undertaken in Namibia and none in Botswana (Vogelsang et al. 2002 and 2010) (Table 1).

Despite the development of anthracology during the last decades, our vision of human-plant interactions remains patchy in this part of the world. Antonites and Antonites (2014), for example, bemourn the lack of anthracological studies concerning Early Farming Bantu communities in South Africa, and their regret applies to the geographical regions and periods of Southern African Prehistory. Majority of the above-mentioned studies concern mainly Pleistocene occupations. Holocene settlements have for various reasons been less in focus.

At most sites from the LSA period, systematic collecting of archaeobotanical remains was not part of excavation priorities. Even when plant remains were collected and analysed, for example at Big Elephant and Geduld in Namibia, and providing interesting results on fruit collection and seasonality, a lack of a systematic sampling of charcoal remains prevented more detailed anatomical and statistical approaches analysis (Wadley, 1979, 2016, Smith et al. 2010, Kinahan 1995, Jacobson 1987). Questions pertaining to past practices of management and exploitation of wood resources as well as the anthropogenic impact on the vegetation are thus difficult to address adequately. For the same reasons, little is known on the distribution of plant species during the Late Pleistocene and the Holocene.

Table 1. Summary of general published charcoal studies in Africa from both the Pleistocene and Holocene periods

Country/region	Site	Period	References
South Africa	Elands Cave Bay	Pleistocene	Allsop 1998
South Africa	Free State	Holocene/Pleistocene	Thornton-Barnett 2013
South Africa	Rose Cottage, Sibudu	Pleistocene	Wadley 1991, 2006, Wadley et al. 2011
South Africa	Sibudu	Pleistocene	Lennox & Bamford 2015, 2016, 2017, Sievers 2006, 2016
South Africa	general	Holocene: farming communities	Antonites and Antonites 2014
South Africa	general	Present	Robbertse et al. 1980
South Africa, Lesotho	Caledon River	Holocene/Pleistocene	Esterhuysen 1996
Lesotho	Western Lesotho	Holocene/Pleistocene	Esterhuysen & Mitchell 1996
Swaziland	Lubombo Mountains	Holocene	Prior & Williams 1985
Namibia	Kaokoland	Holocene/Pleistocene	Eichhorn 2002, Vogelsang et al. 2002, 2010
Mozambique	Chibuene	Holocene: farming communities	Ekblom et al.2014
Cameroon	Dibamba	Holocene: Iron Age	Höhn & Neumann 2017
Central Africa	Muyambe Forest	Holocene	Hubau et al. 2011
Togo	Bassar area	Holocene: Iron Age	Eichhorn & Robion-Brunner 2017
Namibia	Kunene region	Holocene/Pleistocene	In Vogelsang and Eichhorn
Namibia	Northern Namibia	Holocene	Ritcher and Eichhorn 2002
West Africa		Not reported	Rolando 1997

Besides charcoal remains, this study is based on seed/fruit assemblages collected during the excavations at Leopard Cave and Geduld. These types of plant remains can more often be identified at the species level than can charcoal and thus provide a higher taxonomic resolution. Moreover, seed and fruit remains bear witness to other activities than fuel collection, notably foraging for food, fodder and medicinal uses (Thornton-Barnett 2013). In Southern Africa, most fruit/seed studies concern Middle Stone Age sites like Sibudu (Sievers 2015, Sievers 2017, Wadley 2004, 2006),with much less that can be said about Holocene/LSA settlements (Symes 2008, Symes, 2012). From some hunter-gatherer sites in Lesotho there is, for example, mention only of the presence of seed deposits and without further details (Plug 1997).

The objective of this work is thus to produce new data on Late Stone Age environments and plant economies. The botanical identification of charcoal remains from the three sites mentioned above will allow for the reconstruction of the vegetation communities present in the surroundings of the sites during the second half of the Holocene. The perspective is comparative across space and time. Thus, did the populations inhabiting Leopard Cave, Geduld and Toteng benefit from similar environments or did the composition of the flora vary from one site to another? Did the vegetation communities change over time, for example, due to overexploitation and/or climatic change? Can we detect changes in the vegetation that might be linked to changes in the subsistence system such as the introduction of domestic livestock (a question much debated in the context of Southern African Holocene Late Stone Age communities)? Have some species disappeared between this period and today?

Through anthracological analysis, questions can be addressed related to how past populations organised the acquisition of wood for fuel and other uses. Did the past populations visit different habitats for collecting wood? Is it possible to deduce species selection? Were some species avoided as it can sometimes be the case among populations inhabiting these regions today?

Seed/fruit analysis allows approaching several other aspects of past plant economies such as the collecting of plants for food. Through the presence of different species, supplementary information may be gained on the composition of the vegetation cover. Did communities collect fruits and seeds from the same species as those they used for fuel or are there discrepancies between the two datasets? Where were seeds and fruits collected (catchment areas) and when (seasonality)? How were they prepared and consumed?

This book has two introductory chapters (chapters 1 and 2) describing the environmental and archaeological context, while a third chapter is dedicated to the material and methods comprising a section dealing specifically with the constitution of a reference collection; an indispensable tool for the identification of archaeological charcoal samples. The anatomy of the wood species making up the comparative collection is presented in detail as an appendix, in the form of plates and descriptions. Chapter four presents the results of first-hand ethnographical field research in the form of interviews and surveys on plant exploitation and use conducted among present-day local Damara and San communities. In chapter five, the results of identifications of wood charcoal and seed/fruit remains from Leopard Cave, Geduld and Toteng are presented. The taxa identified are described in detail, followed by a presentation of the quantitative results from each site. The discussion in terms of paleoenvironment and plant uses constitutes chapter 6 followed by a conclusion and future perspectives. The latter addresses the actions that can be employed in advancing archaeobotanical studies in southern Africa, including expanding the study to neighbouring sites and beefing up the reference collection.

1

Environmental Background of Southern Africa

1.1. Delimination of the study area

The three sites concerned by this study are located in the west/central parts of Southern Africa, slightly north of the Tropic of Capricorn (or Southern Tropic) and south of the vast alluvial fan created by the endorheic delta of the Okavango River in northern Botswana (Figure 2) The study zone is comprised within the semi-arid Kgalagadi[4] basin that extends over 2.5 million km^2 in Southern Africa mainly in present-day Namibia, Botswana and South Africa with minor encroachments also into Angola, Zambia and Zimbabwe in its northern parts.

The Kgalagadi (also Kalahari) desert, covering a surface of approximately 900,000 km^2 constitutes the heart of a homonymous basin. The present-day Kgalagadi basin is characterised by a diversity of open landscapes comprising savannahs, shrublands, grasslands and sand deserts (Van Wyk and Van Wyk 2007).

1.1.1. Geomorphology and soils

During the late Jurassic and early Cretaceous, Southern Africa underwent a number of significant epirogenic[5] movements (Tooth 2002). Recent investigations of proxy evidence for climatic and environmental changes (Table 2) suggest that a temporally and spatially complex succession of events including a complex record of punctuated aridity, dunefields and aeolian deposits characterised the evolution of Southern Africa during the late Quaternary period (Shaw 2003; Thomas & Shaw, 2002). Multi-proxy data including lacustrine carbonates, diatoms, shoreline sediments, pollen records etc. has allowed the reconstruction of landforms, climate and biodiversity over a long period covering the late Pleistocene and the Holocene periods. Aeolian records reflect an arid phase that occurred during the Last Glacial Maximum(LGM) while the beginning of the Holocene presents warmer and more humid conditions due to sea surface temperature (SST) movements (Humphries et al. 2017). Such variations in the climate, of course, had considerable influence on the flora and fauna of the region. Palaeoenvironmental research is continuing in Southern Africa with the quest to further elucidate the longstanding history of past climatic changes as well as to comprehend possible changes.

Geomorphological data details the Kgalagadi basin's geological past and its evolution during the Quaternary period (Burrough et al. 2009, Matmon et al. 2015, Shaw 1988). The basin is mostly flat and comprises one of the largest sand bodies in the world. The middle Kgalagadi, in northern Botswana, includes three paleolake basins: Makgadikgadi, Ngami and Mababe. They constitute a vast area of internal drainage where ancient aeolian, fluvial and lacustrine sediments can be traced (Burrough and Thomas 2008). The paleolakes are associated with the still prevailing Okavango Delta and together they cover a total area of more than 60 000 km2. The abundance of ancient fluvial and lacustrine shorelines has rendered this area highly interesting for paleolacustrine research for more than four decades (Shaw 1985). Thomas et al. (2003) have documented the geomorphological setup of the Middle Kgalagadi especially the Lake Ngami, establishing a network of the relic and current inflows substantiated by remnants of ancient shorelines. The dating of deposits using diatom records and luminescence dating has revealed Holocene modifications of older landforms as well as fluctuations of the lakes' surfaces and levels during the Holocene period. Similar studies carried out in the Tsodilo hills used lacustrine sediments and carbonates, molluscs and diatoms as well as Accelerated Mass Spectometry (AMS), Optically-Stimulated Luminescence (OSL) and carbon-14 dating (^{14}C) (Thomas et al. 2003). This study's multi-proxy approach has determined the timing of fluctuations between wetter and drier Quaternary periods in the Kgalagadi Basin. However, the approach is limited to certain areas of the basin, as aeolian and lacustrine formations in stratigraphic developments are not found everywhere.

The northwestern part of the Kgalagadi basin is made up of remnants of aeolian dunes and ephemeral springs and drainage ways (Burrough and Thomas 2008). Most of Southern Africa consists of plateaus covered by soils that can be as old as a million years (Dewitte et al., 2013). The determinants of the soil patterns show differences in age, parent material, physiography and past and present climatic conditions. The majority of soils found in the region are sandy arenosols composed mainly of quartz particles with poor water holding capacity and poor nutrient contents (Figure 3). Wind-blown sands cover large tracts of the south-central and southeastern parts of the subcontinent and are predominant in present-day Botswana. These loose sandy deposits result from the drying up of the Kgalagadi group of paleolakes as well as other fluvio-lacustrine formations.

To the southwest and further south, the region is made up of xerosols, regosols and yermosols which are all

[4] Kalahari was derived from the word Kgalagadi which refers to "the land that has dried up" or "the land of thirst". The name stems from the Setswana verb "*go kgala*" meaning to "dry up". The words Kalahari & Kgalagadi have been used interchangeably in literature over the years.

[5] Vertical tectonic movement that compresses the earth's crust in slow displacements usually showing long wavelengths but exhibiting little folding, rather tending to lift whole land masses evenly.

Figure 2. Map of Southern Africa with the localisation of the study sites and geographical features mentioned in the text. Map author: David Pleurdeau.

Table 2. Summary of the current geomorphological literature bearing on Southern Africa

Proxy and period	Region	References
Present (1948–2000)		
Rainfall and temperature	S and W Africa	New et al., 2006
Meteorology, sea surface temperatures	SW coast of Africa	Nicholson & Entekhabi1987
Holocene		
Dune deposits	SW Kgalagadi	Blumel et al. 1998
Pollen from cow dung	Botswana, Zimbabwe, South Africa	Carrión et al. 2000
Archaeology	Kgalagadi	Denbow, 1986
Fluvial deposits, soil sediments	NW Namibia	Eitel et al. 2002; 2005; 2006
Pollen from hyrax middens	NW Namibia	Gil-Romera et al 2007a, 2007b
Dunes, marine and fluvial sediments	Namibia	Heine, 2004; 2005
Stalagmites	NE South Africa	Holmgren et al., 1999
Aeolian sediments	E coast Southern Africa	Humphries et al. 2017
Lacustrine sediments, diatoms	Lake Ngami basin, NW Botswana	Huntsman-Mapila et al. 2006
Plant biomarkers, pollen records	N Angola and central South Africa	Rommerskirchen et al., 2006
Palaeoclimatic data	Southern Africa	Tyson & Lindesay, 1992
Stalagmites	NE South Africa	Tyson et al. 2002
Stalagmites	NE Namibia	Voarintsoa et al. 2016
Stalagmites	N Namibia,	Sletten et al., 2013
Lacustrine archives	S Cape	Wündish et al 2016
Pliocene-Holocene		
Calcretes/Carbonates	Kgalagadi	Watts, 1980
Pleistocene-Holocene		
Lake beach ridges	Northern Botswana	Burrough et al., 2007
Calcrete, fossil wood	Middle Kgalagadi	Shaw & Cooke., 1986)
Carbon isotopes from bovid teeth	N Cape province, South Africa	Lee-Thorp & Beaumont, 1995
Pollen from hyrax dung	NW Namibia	Scott et al. 2004
Calcrete, fossil wood	Middle Kgalagadi	Shaw & Cooke., 1986)

Proxy and period	Region	References
	Late Quaternary	
Temperature, rainfall	South Africa	Chevalier& Chase, 2016
Lacustrine and fluvial features	Middle Kgalagadi	Burrough et al., 2009
Pollen from hyrax middens	South Africa	Gil-Romera et al., 2006
	Pleistocene	
Pollen	Southern Africa	Scott and Neumann 2018
Palaeoclimatic data	Africa	Blome et al. 2012
Pollen	South Africa	Meadows et al.,1996
Palaeoclimatic data	Namib desert	Lancaster, 2002
Pollen from hyrax dung	SW Cape, South Africa	(E. M. Scott, 1994)
Deep-sea sediments	NW Africa	Stuut, et al., 2004)
Cave speleothems	Cape Province, South Africa	Talma & Vogel, 1992
Meteorological records	W coast Southern Africa	Stuut & Lamy, 2004
Diatoms, sand ridges, lake sediment	NW Kgalagadi	Shaw et al., 2003)
Silt deposits	Central Namib desert	Srivastava et al., 2006
Lacustrine formations.	NW Botswana	Shaw, 1985; 1988
Sedimentary cycles	Kgalagadi basin	Matmon et al., 2015
Diatoms, aeolian deposits, speleothems	Northwestern Kgalagadi	Thomas et al.2002
Wetlands, wetland deposits	Eastern Free State, South Africa	Tooth and McCarthy 2007
	Other periods	
Sediments, paleosoils, in situ fossils	Karoo, South Africa	Smith et al.,1995
Alluvial ridges	Eastern Free State, South Africa	Tooth et al. 2002

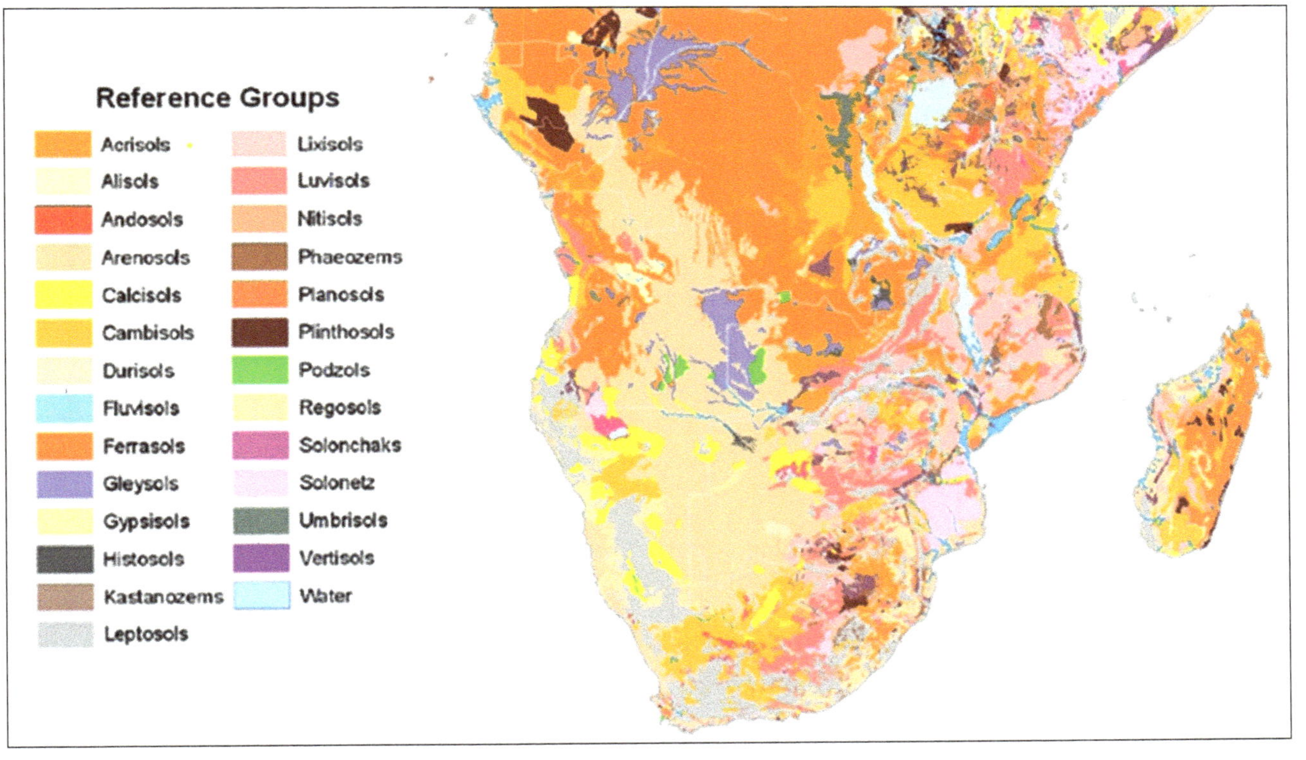

Figure 3. Major soil types of Southern Africa. Data source: Dewitte, O., Jones, A., Spaargaren, O., Breuning-Madsen, H., Brossard, M., Dampha, A., Deckers, J., Gallali, T., Hallett, S., Jones, R., Kilasara, M., Le Roux, P., Micheli, E., Montanarella, L., Thiombiano, L., Van Ranst, E., Yemefack, M., Zougmore, R., 2013. Harmonisation of the soil map of Africa at the continental scale. Geoderma 211–212, 138–153. https://doi.org/10.1016/j.geoderma.2013.07.007. Map author: Stefania Merlo.

aridisols and in conformity with the arid and desertic coastline area of western Namibia. Fluvisols are present in areas that have been flooded by surface waters or rising groundwater, from rivers or delta formations. In the north-central parts of the region, in southern Zambia, the soils are mostly of ferralsol type, with low occurrences of arenosols. Ferralsols are old soils that have developed on geologically old parent material. A more diversified situation is visible in the interior of the Mozambican coast to the east and towards the eastern parts of the subcontinent in South Africa's Eastern Cape. Here a mixture of luvisols, vertisols and arenosols and planosols are present. The diverse landforms of this area with mountains, lowlands and wetlands contribute to a higher diversity of soil types. In the west (Namib Desert), the dominant soils are arenosols, gypsisols and leptosols together with sand dunes, gravel and rock outcrops (Heine 2005).

1.2. Past and present climatic conditions

The study region belongs to the Semi-Arid and Arid Zones of the African continent, as defined by Bationo et al. (2012). The Semi-Arid Zone is characterised by average annual rainfalls of 200–800 mm and is located between the sub-humid wooded savannah zone and the Arid Zone, at latitudes ranging between 15°–25°S (Figure 4). The Arid Zone extends north of latitude 20°S and south of latitude 30°S with annual rainfalls not exceeding 200 mm. The immense Saharan Desert in North Africa and the Namib and Kgalagadi deserts in Southern Africa are part of this arid zone. During the Last Glacial Maximum (LGM), ice sheets covered much of North America and northern Eurasia, affecting climates worldwide by initiating drought, desertification and a dramatic drop in sea levels and temperatures. Most records show that generally dry conditions associated with lower tropical land and sea surface temperatures were common in both hemispheres (Chase 2000). In Africa and the Middle East, smaller mountain glaciers formed as sandy deserts expanded (Thornton-Barnett 2013). In equatorial and North Africa, two drastic humid/arid transitions believed to match the major Greenland warming events associated with the switching of the oceanic thermohaline circulation to the modern mode that occurred circa 1514.5 and 11.511 ka cal. BP (Chase 2000). In southern Africa, a period of fluctuations between wet and dry conditions was observed with patterns of drier conditions in the eastern and some western sides of Southern Africa and wetter conditions in the southwest (Stone 2014). Nkoana et al. (2015), report that during the LGM, winters in most of Southern Africa were significantly wetter than they are today.

Evidence from a compilation of pollen data from Southern Africa indicates that colder temperatures prevailed during the LGM (Scott et al. 2012). In the northern Namib desert, pollen data covering the last 21 000 years also shows that this region was characterised by colder and more arid conditions than at present with a vegetation cover formed by lower elements from the Asteroideae, Ericaceae, and Restionaceae groups. The Late Glacial to early Holocene transition saw two arid periods, between 14.4–12.5 and 10.9–9.3 ka cal. BP, a period that can be linked to the Younger Dryas. Marine pollen taxa from the Mozambique Channel (Elmoutaki, 1994) similarly indicates that the climate was drier than today in eastern Southern Africa during the late Pleistocene (Younger Dryas).In the early Holocene, rainfall increased along the coasts due to the warming of the Indian Ocean while the central interior remained dry until circa 7000 cal BP. The middle Holocene remained generally wet but variable according to the regions. The warmest and most humid period in the Holocene was registered between 6300 and 4800 cal. BP and was then followed by a first dry spell. Cooler conditions occured around 3 ka BP and more uniform dry conditions are recorded across the subcontinent from around 2 ka BP (Scott and Neumann 2018).

Inferred warmer and more humid conditions at the start of the Holocene correspond to an extended phase of increased SST around southern Africa. This interpretation is consistent with southern African temperature records, which generally show that interglacial temperatures were established by 11000–10000 cal BP, with maximum warming, reached during the Holocene Altithermal at 8000–6000 cal BP. The Mfabeni pollen record shows forest growth and expansion (Finch and Hill, 2008) indicating warm and relatively moist conditions during the early Holocene (Humphries et al. 2017). More favourable conditions in the Early Holocene followed by a drier Middle Holocene is also purported by the investigations conducted by Scott and Woodborne (2007a).

The past Southern African climates and their driving factors, especially in the Holocene, are still incompletely understood. This is due to the absence of proxy records, and their scarcity in connection with preservation, because predominantly arid conditions provoke the erosion of certain deposits. Despite their rare occurrences, significant environmental changes in the Late Holocene can still be traced. In the southern Cape, for example, the changes are significant over the past 4200 years. Data from geo-archives show fluctuating lake levels suggesting a drier environment than now, with wetter conditions because of heavy rainfall that occurred between 3760–3690 cal BP (Wündsch et al., 2016). Similarly, the period between 2710 and 700 cal BP until the past 140 years the climate was wetter in comparison to today (Bousman & Brink, 2018).

Pollen records from the Tswaing crater in South Africa show both climate and vegetation fluctuations in the south-central parts of South Africa in the Early to Late Holocene (Metwally 2014). Drier and warmer climatic conditions

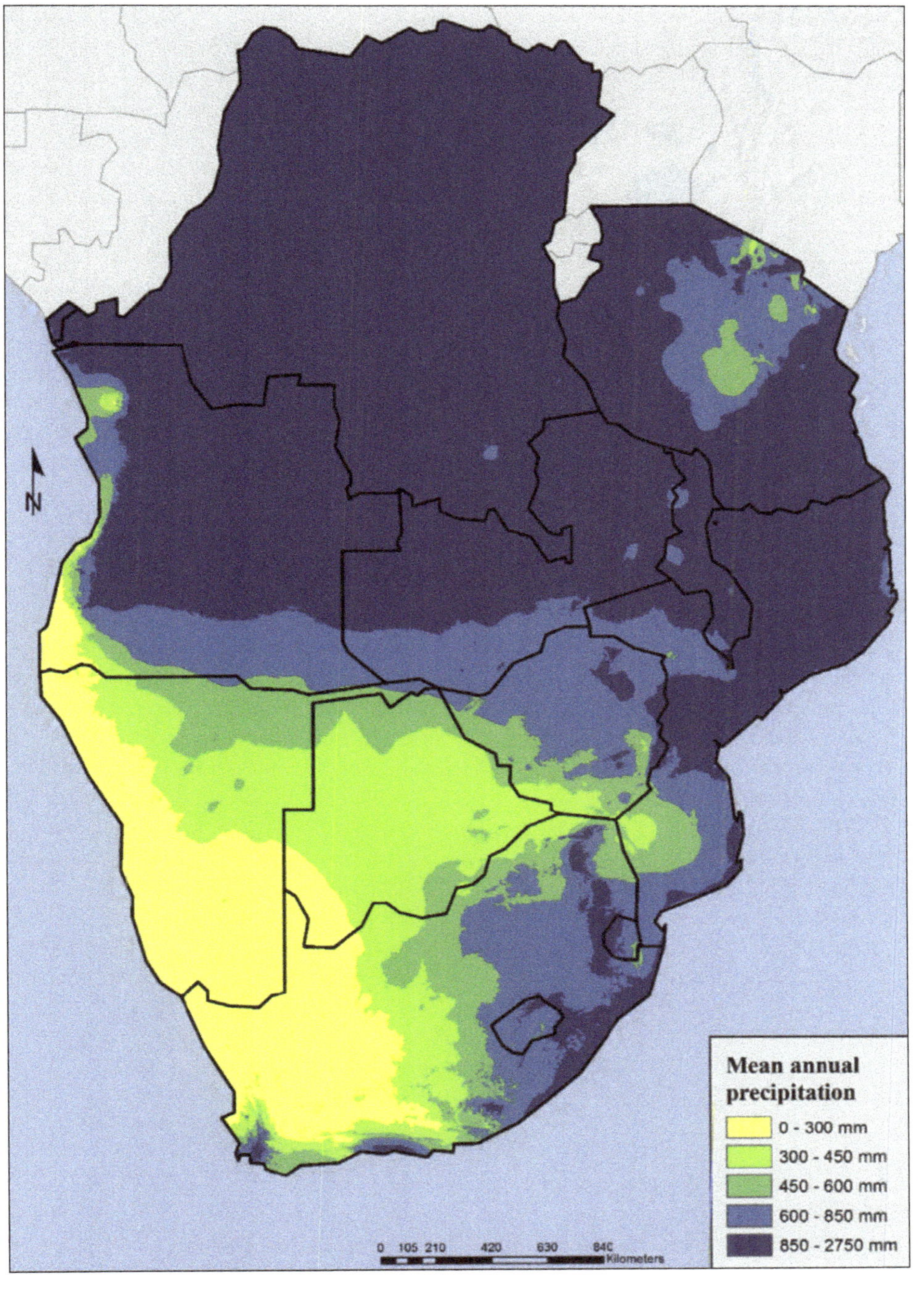

Figure 4. Current annual precipitation over southern Africa (after Mafongoya et al., 2017).

were recorded from the Early to the Mid-Holocene, while warmer and wetter conditions occurred in the mid-Holocene in the periods around 7000–1800 cal BP, with a dry hiatus between 3800–3600 cal BP.

The highest rates of aeolian sedimentation are coincident with the Late Glacial Maximum, a period of increased aridity across a great area of Southern Africa 11 500–7900 BP. (Truc et al. 2013; Chevalier and Chase 2016). The faunal and charcoal assemblages from Elands Bay Cave suggest that conditions remained relatively moist in the southwestern Cape throughout the Pleistocene/Holocene transition, 7900–5300 cal BP. This period was the warmest of the last 10,000 years at several sites across the subcontinent (Mitchell 1997). Pollen sequences in the Mpumalanga bushveld suggest a greater prominence of broad-leaved savannah species (and thus wetter conditions) by 7500 cal BP.

Stalagmites can store multiple climatic proxies in the form of stable isotopes, petrographic and mineral indications that vary according to environmental factors such as solar activity and temperature changes (Voarintsoa et al., 2017). A stalagmite record from the Dante Cave in northeastern

Namibia shows a gradual transition from wetter to drier conditions from 4600 to 3300 cal BP followed by a return to wetter conditions at around 1800 cal BP. The most recent wet period during which further wet/dry conditions prevailed has been recorded at around 230–100 BP. Apparently, the northern part of the Kgalagadi desert was equally wet around the same time with the Makgadikgadi paleolake, a crucial source of water. (Sletten et al. 2013, Pleurdeau et al. 2012).

Like other continents, the African landmass has undergone important climatic fluctuations during the Quaternary period. Still, as most of the continent lies in the tropics, it has been less affected than the northern hemisphere by glacial activities during the Pleistocene. Instead, the pivotal influences of climate change have been due to variations in precipitation, evaporation and runoff patterns, induced by the recurrent latitudinal movement of the Inter-Tropical Convergence Zone (ITCZ)[6] as well as oceanic current, thermohaline circulation and current air masses (Nicholson 2000).

Today, only a small part of Southern Africa is humid, i.e., the zone around the southern and eastern extremities. The interior and the western margins are mostly dry with annual potential evaporation greatly exceeding precipitation. The arid conditions in Namibia and Angola are a result of the cold winds on the west coast as opposed to the moisture-rich air to the southeastern and eastern coasts, resulting in precipitation over the mountainous areas of the Drakensbug where mean annual rainfall may attain 2000 mm. Mean temperatures vary according to the season and the region and range in the summer from 18° in the mountains to around 40° elsewhere. The degree of aridity generally decreases along an east-west gradient, from sub-humid and semi-arid in the eastern interior to arid and hyper-arid in the west (Tooth and McCarthy 2007).

The interior landmass of Southern Africa is characterised by three major rainfall zones, namely the summer rainfall zones due to the movements of the ITCZ leading to the tropical easterly trade winds (Wündsch et al. 2016). The ITCZ migrates south during the southern hemisphere summer, divides the ridge of high pressure responsible for the stable winter drought conditions (April-October), and allows the penetration of the Indian Ocean easterly low pressure bringing rainfall into the southern summer rainfall zone (Burrough et al. 2009). Likewise, the winter rainfall zone lies along the west coast and the westerly trade winds serve as the chief source of precipitation. Lastly, the Year-round Rainfall zone represents the transition between the two latter zones and enjoys summer and winter rains. The Atlantic Ocean is a secondary source of moisture for Southern Africa also strongly associated with the ITCZ and supplies moisture via the tropical westerlies (recurved equatorial trade winds) principally to the region northwest of the Middle Kgalagadi at approximately 12°S. Angola, the source region of the Middle Kgalagadi's principal river systems, marks the southernmost limit of the rain-bearing systems associated with the West African Monsoon. Ocean currents largely determine shifts in atmospheric circulation and changes in sea surface temperatures (Humphries et al. 2017, Wündsch et al. 2016). The two main currents are the warm Algulhas in the Indian Ocean, influencing the southern and eastern parts of the mainland, and the cold Benguela current of the Atlantic Ocean to the west (Voarintsoa et al., 2017). Voarintsoa et al. (2017) also states that climate changes in northeastern Namibia depend on solar action and that changes in global temperatures probably have resulted in longer lasting summer rainfall seasons. In the same region, cooler conditions were recorded during the periods between 1660–1710 and 1790–1830 and wetter periods comprising longer summer seasons were linked to the progress of ITCZ towards the southwest. A slight advance of the Angola–Benguela Front (ABF) to the south was frequent at the time, bringing more rainfall inland. On the other hand, drier and warmer periods in northeastern Namibia, inferred from stalagmites post CE 1715, correspond to globally warmer climatic conditions, suggesting a migration of the ITCZ northeastwards, specifically with more warming of the Northern Hemisphere. This is concomitant with decreasing rainfall in the summer rainfall zone of southern Africa since ca. CE 1900.

As a result, the Southern African climate is diverse, ranging from the warm and wet conditions in the eastern part to semi- to hyper-arid climates in the central and western areas. Mediterranean climate conditions also prevail in the southwest.

1.3. Past and present vegetation

During most of the Holocene period water was available, temperatures increased, and the Desert biome expanded in some areas (Lim et al. 2016). Over time from around 16 000 to 10 000 cal BP the ratio of pollen from the Poaceae family compared to that of the Asteraceae family increased, which may be interpreted as an increase in moisture in the open, low-lying Nama-Karoo biomes found in southern Namibia and South Africa. Increased precipitation may have caused the Nama-Karoo on the southern side to expand into the comparatively grassier Grassland biome towards central and southeastern South Africa. Despite being dominated by semi-arid to arid conditions, the flora of present-day Southern Africa comprises more than 24 000 known plant species (Gibbs Russell, 1987). Vegetation markers show a generally dry early and middle Holocene in Southern Africa. The results from the Western Cape's Pakhuis Pass pollen sequence suggest that the Early and Middle Holocene periods were much drier than today (Scott & Woodborne, 2007a). The dry spell is in part caused by an ITCZ system that encouraged dry air zones

[6] The strong seasonality with summer monsoonal rainfall characterised the climate of a large part of Africa influenced by the recurrent movement from North to South of the ITCZ during the year.

currently located in the Karoo biome. At around 8 ka BP, C4 adapted grasses are found supporting this attestation. The period between 5 and 2 ka BP shows the prevalence of more succulent vegetation types, suggesting a cooler period. This conclusion is supported by Gil-Romera et al. (2007) through pollen records depicting vegetation conditions and changes in the north-western margin of Southern Africa from 6 ka BP to the present and major changes towards the Later Holocene where the vegetation shifted to an increase in the tree/grass proportions circa 4.2 ka BP and 1.6 ka BP resulting in arid savannahs. Prolonged periods of rain encouraged the expansion of savannah woodlands, mostly *Colophospermum mopane* and Combretaceae species (Eitel et al. 2002). A vast area of the southern African highland plateau consists of at least some woodland savannah with *Brachystegia* and *Julbernadia* species that often coincide with granitic basement rock. Mid-Holocene sediments and pedological studies revealing vertisols, kastanozems and calcisols on the surface suggest they were formed under open grassland conditions in the northern parts of Namibia. Arid conditions were also recorded in southwestern Africa in northern Kgalagadi and the northern Namib desert towards 10k BP (Shp et al. 1998). The warmest and most humid conditions prevailed around 6–4 ka BP. The last 2000 years of the Holocene have experienced an increasing anthropogenic impact on the vegetation cover.

Today different models are used to contextualise biomes (Figure 5) (Scott and Neumann 2018). First described biomes are viewed as a community of vegetation displaying similar vegetation characteristics or as a land community at a given geographical scale, whether continental and sub-continental level but also incorporating the faunal components among other various life forms present. But depending on the question at hand, different authors treat the definition of biomes as including the taphonomic characteristics they are interested in, for example, considering biomes solely by botanical characteristics. Thus, Rutherford et al. (2006) describe biomes as "a high-level hierarchical (hence simplified) unit having a similar vegetation structure exposed to similar macroclimatic patterns, often linked to characteristic levels of disturbance such as grazing and fire." Southern African biomes consist of a large variety of rich biomes and this chapter will be used to take into consideration the initial description of the term.

The southern African biomes are currently recognised as composed of (cf Table 3):

- the mesic Forests Fynbos with poor soils,
- Grassland and Savanna with average rainfall ranging from 200–3000mm,
- and the more arid Indian Ocean Coastal Belt Nama-Karoo, Succulent Karoo and Albany Thicket Desert with rainfall levels ranging from 10–520 mm at least in the case of the Nama-Karoo (Truc et al., 2013). These are divided into summer rainfall or winter rainfall biomes. The two biomes, namely Albany Thicket Biome which Rutherford & Westfall (1994) previously referred to as unmappable 'dwarf forest' of the Eastern Cape and the Indian Ocean Coastal Belt off to the Eastern coastal cape bordering the Savanna biome in the mainland between the border of South Africa and Mozambique were omitted from earlier literature, while (Rutherford et al., 2006) included them in recent classification.

The wetter winter-rainfall regions, which include the Fynbos biome to the west, the Succulent Karoo biome of the Namaqualand together with the Little Karoo (the

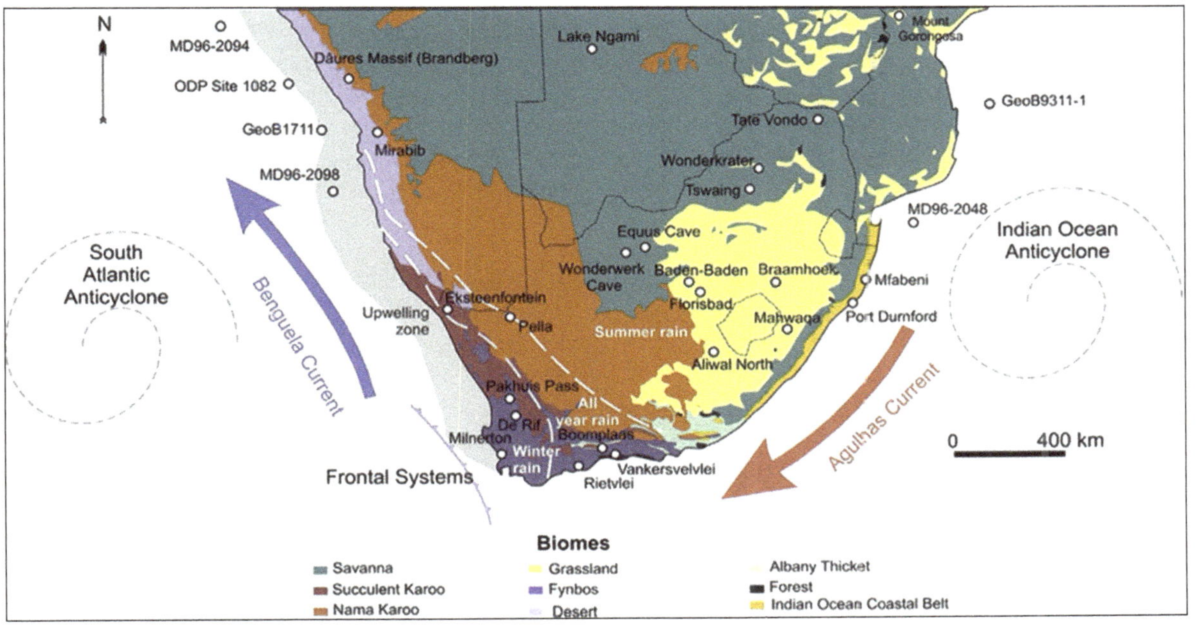

Figure 5. Biomes of Southern Africa (after Scott and Neumann 2018).

Table 3. Physiognomy and main taxa of Southern African biomes

Biomes (Russel 1987, Carrión, 2000)	Biomes (Rutherford et al 2006)	Vegetal cover dominance	Characteristic taxa
	Summer rainfall biomes		
Desert	Desert	Mostly ephemeral and annual plants	Poaceae Cucurbitaceae Asteraceae
Nama Karoo	Nama-Karoo	Short, open low lying shrublands	Poaceae Opuntia aurantia Prosopis glandulosa(invasive species)
Savannah	Savannah	Woody C4 grasslands	Poaceae
Grassland	Grassland	Grass cover with little or no trees	Poaceae
	Winter rainfall biomes		
Fynbos	Fynbos	Renosterveld and heathland	Restionaceae, Ericaceae, Asteraceae, Bruniaceae, Fabaceae, Rhamnaceae, Thymelaeaceae
The succulent Karoo	Succulent Karoo	Dwarf succulent dwarfs with rare C3 grasses	Crassulaceae Mesembryanthemaceae
	Indian Ocean Coastal Belt	Forest cover, grassland, dense savanna	
Albany Thicket	Dense spiny and succulent vegetation		

most species-rich semi-desert on earth), offer the highest assortment of succulent plants in the world. The Savannah biome, to the north and east areas of summer-rainfall calibre, is the largest biome in Southern Africa. The Grassland biome that is also of summer-rainfall type is also rich in species (Rutherford et al 2006).

Some areas of Southern Africa covering shallow water wetlands are commonly associated with a specialised range of freshwater and flood tolerant woody species like *Syzygium* and palms. Wetlands are distributed from the top of the catchment to the deltas at the bottom, indicating clearly the networked interdependence of water resources. Wetlands are often indicators of substantive groundwater resources and fluvial buffering, in particular, important in sediment and nutrient trapping (Scott & Neumann, 2018).

The Mozambican coastal area is made up of mangrove forests while herbaceous swamps with papyrus and reeds are also widespread and common. There are relatively small areas of high-altitude swamps, bogs and mires, which are characterised by Afro-alpine vegetation which is specialised and often endemic. Open water may have potamogetons and is often colonised by floating *Pistia* while *Echinochloa scabra* is common in the shallow waters. Clearwater rivers downstream of wetlands such as the Okavango Delta contain much-submerged vegetation, including water lilies *(Nymphaea* sp.), the tubers of which are often harvested for food. Most wetland plant species have some economic value, as food or more usually as thatching or screen or carpet-making material (Taylor et al., 1995).

The south-western corner of Namibia on the other hand, is at present a semi-desert environment, with mean annual precipitation of less than 100 mm (Vogelsang et al., 2010). The low precipitation rates strongly influence the vegetation, which is a Dwarf Shrub Savannah, characterised by low shrubs and grassland that are drought resistant.

1.4. Fauna

Apart from Africa, the Pleistocene is marked by the extinction of mega-fauna. In Africa and especially in sub-Saharan Africa a diverse fauna has persisted into current times (Table 4). The overkill hypothesis (Wroe et al., 2004; Smith, 1998; (Surovell et al., 2005) Wells, 1969) does not seem to apply here, rather a number of ecological modifications during the Pleistocene. Faunal diversity is more developed in certain environments than others, such as in, the Southern Cape with more sclerophyllous vegetation. Quasi-no grassland adapted for grazers hosts a comparatively lower number of species and fewer animals

(Deacon, 2017). Faunal species distribution, spread, and extinctions may be due to a number of factors, most importantly as a response to climate change. The major climatic and environmental changes southern Africa underwent during the Tertiary and the Pleistocene periods that provoked vegetation changes could also have possibly influenced the distribution of fauna with species diversity resultantly high. Archaeologically, research has been able to trace the evolution and distribution of southern African fauna, especially during the Pleistocene through to the Holocene (Erasmus et al., 2002). Since certain species conform only to specific environmental conditions, faunal data can be used to deduce the environmental changes and vice-versa.

Table 4. Faunal taxa currently found in Southern Africa

Faunal classes	Faunal description	Charasteristic family examples
	Mammals	
Cloven-hoofed mammals		Bovideae
Bats		Giraffidae
		Pteropodidae
		Rhinolophidae
		Elephantidae
		Chrysochloridae
Moles		Macroscelididae
Hyraxes		Procaviidae
Rats, mice	**Grazers and browsers, rodents and meat-eating predators**	Muridae
Hares, rabbits		Leporidae
		Hyaenidae
Cats		Felideae
		Canidae
		Mustelidae
Sealions, seals		Otariidae
Horses		Equidae
Suids		Suidae
	Birds	
		Spheniscidae
		Podicipedidae
		Diomedeidae
		Procellariidae
		Ardeidae
	Terrestrial and freshwater birds	Ciconiidae
		Anatidae
		Pandionidae
		Otididae
		Dromadidae
		Charadriidae
	Reptiles	
Crocodiles		Crocodylinae
		Colubridae
Snakes		Elapidae
Chameleons		Pythonideae
Turtles	**Both teresstrial and aquatic reptiles**	Chamaeleonidae
		Pelomedusidae
		Agamidae
Lizards		Cordylidae
		Scincidae

(*Continued*)

Table 4. Faunal taxa currently found in Southern Africa (*Continued*)

Faunal classes	Faunal description	Charasteristic family examples
Fish		
Coastal fish	Marine and freshwater fish	Myxinidae
		Gymnuridae
		Pristidae
		Rhinobatidae
		Narkidae
		Hemigaleidae
Fresh water fish		Clariidae
		Cyprinidae
		Gobiidae
		Cichlidae
		Galaxiinae
Amphibians		
Toads Rain frog	Frogs and toads	Bufonideae
		Brevicipitidae
Frogs		Phrynobatrachidae
		Rhacophoridae
Insects		
	Grasshoppers, flies, dragonflies , mosquitoes and beettles	Tettigoniidae
		Synlestidae
		Lucanidae
		Carabidae
		Lycaeninae
		Sphingidae
		Pyrrhocoridae
		Culicidae
		Glossinidae
		Saturniidae
		Acrididae
Arachnids	**All spiders**	Araneidae
Snails		
	Both marine and terrestrial gastropods	Ampullariidae
		Assimineidae
		Ellobiidae
		Littorinidae
		Lymnaeidae
		Neritidae
		Physidae
		Planorbidae
		Potamididae
		Succineidae
		Thiaridae
		Viviparidae

2

The Late Stone Age with a Special Focus on the Studied Sites

2.1. The Late Pleistocene and the onset of the LSA in Southern Africa

The transition from the Pleistocene period to the onset of the Holocene is accepted to have taken place around 11, 000 cal. BP (Stynder et al. 2007). The Younger Dryas is the longest and coldest of several very abrupt climatic changes that occurred at the end of the Late Pleistocene (Easterbrook 2012) into the present interglacial Holocene. It is a very important period to climatologists and geologists when it comes to understanding climatic changes and global warming. The Younger Dryas stadial was a brief geological period characterised by extreme cold climatic conditions and drought, which occurred between approximately 12,800 and 11,500 BP (Harrison and Glasser 2011, Currant and Jacobi 2011). The changes preceding the Younger Dryas include sudden global warming 14 500 BP that caused the immense Pleistocene ice sheets to retreat, followed by fluctuating cool and warm conditions between 14 000 and 11, 500 BP. At the end of the Younger Dryas, ca. 11 500 BP, the rise of temperatures and humidity resumed and the early Holocene warming period is believed to have caused the melting of the North American and European ice sheets sending a surge of fresh water into the North Atlantic. Thermally fractionated nitrogen and argon isotope data from Greenland ice core GISP2 indicate that the summit of Greenland was approximately 15 °C colder during the Younger Dryas than today (Alley 2000; Thornton Barnett, 2013).

Even though a few Middle Stone Age industries extend into the Holocene period in Southern Africa, the Late Stone Age (LSA) is the most dominant. The term Late Stone Age was first coined by Goodwin in the 1920s (Goodwin, 1926; 1929) to designate the Early LSA industries that started around 45, 000 cal. BP and lasted until the Last Glacial Maximum, originating in the eastern parts of Southern Africa and spreading westward (Scott & Neumann, 2018). The material culture relating broadly to this time range has been used to distinguish it from other periods.

The transition between the Middle and the Late Stone Ages is thought to be related to the introduction of the southern and eastern African Magosian industries that took place between the beginning of the Holocene to ca. 8000 cal. BP (Deacon 2017; Clark 1959, Mcbrearty and Brooks 2000). The appearance of blade and complex stone tool techniques were perceived as a reference point for the evolution of the microlithic LSA industries going as far back as 40 000 BP (Table 5) with incipient techniques such as the bipolar technique as evidenced from the Border cave (Villa et al. 2016). This definition was inadequate, because the LSA was not exclusively microlithic. Instead of understanding the LSA strictly through of microlithic industries, it is better to view its microlithics as a late lithic industry built within a framework of lithic adaptation (Clark 1959; Deacon 2017).

The inhabitants of the MSA created the first art and symbolic culture. They created arrows, worked hide and had working tools hafted with unique glueing techniques. The lithic technology included Levalloisprepared core blades and composite microliths. MSA traditional cultural industries include pre-Still Bay, Still Bay, Howiesons Poort, and post-Howiesons Poort. The Pre-Still Bay lithic industry is the earliest of the MSA characterised by Levallois cores, unifacial points and large blades and

Table 5. Summary of current chronological frame of MSA and LSA in Southern Africa

MIS dates	Chronozone	Cultural period	Industry/Culture	Beginning
MIS 1–11 ka BP	Holocene	Later Stone Age	Smithfield Wilton Oakhurst	860 ~7,8ka BP?
MIS 2–24ka	Late Pleistocene		Oakhurst Robberg	~13ka
MIS 3–58ka			Robberg Early LSA	~26ka ~48ka
MIS 4–73ka		Middle Stone Age	Post HP MSA Howieson's Poort Still Bay Pre-Still Bay MSA	59 500ka ~65ka ~71ka
MIS 5–129ka			MSA	From ~280ka

flakes assemblages. The Still Bay (SB) is another of the MSA lithic industries named after the Stillbay Site where it was first described (Goodwin, datings to between 71,900 and 71,000 BP) (Jacobs et al. 2008). The Still Bay style of lithic manufacture is distinctive for its association with bifacial points. The Howiesons Poort (HP) lithic industry was named after the Howieson's Poort Shelter near Grahamstown (Goodwin and van Riet Lowe 1929). The HP industry lasted for a period of about 5000 years, from 65,800 to 59,500 BP (Jacobs et al. 2008) and is distinctive for including both modern and pre-modern lithic assemblages.

In its initial stages, the LSA was interpreted as a technological phase representing a natural unit that included the two significant cultures of Wilton and Smithfield (Goodwin & van Riet Lowe, 1929). During this period a sophisticated microlithic industry appeared circa 12 000 cal. BP only later to revert to a macrolithic then circa 8000 cal. BP shift back to a microlithic one associated with the hunting of small mammals. Today, many other archaeological sites in Southern Africa document the heterogeneity of the LSA material culture (Figure 6).

The period was also later linked to the appearance of the modern hunter-gatherer San communities of southern Africa. The problem, however, lay in that they were interpreted using as a framework the contacts between the Mesolithic and Neolithic cultures (Deacon, 2017); LSA sites are not easy to compare with other regional cultures that occurred contemporarily.

2.1.1. Technical changes during the LSA

The LSA period coincides with a shift in lithic technology, raw material uses and subsistence approaches. Though large number of sites (both shelter and cave sites) exist in Southern Africa, few have provided a clear chronology dating back to the Late-Terminal Pleistocene. Numerous Late Stone Age traditions in Africa appeared in the Late Pleistocene (Plug 1997) approximately between 40 000 and 12 000 BP and are known to be linked to local adaptations (Figure 7) Their technical distinction is mainly based on their degree of microlithization and on their bladelet components.

When the concept of Late Stone Age was first coined in the 1920s (Goodwin 1928; Goodwin and van Riet Lowe

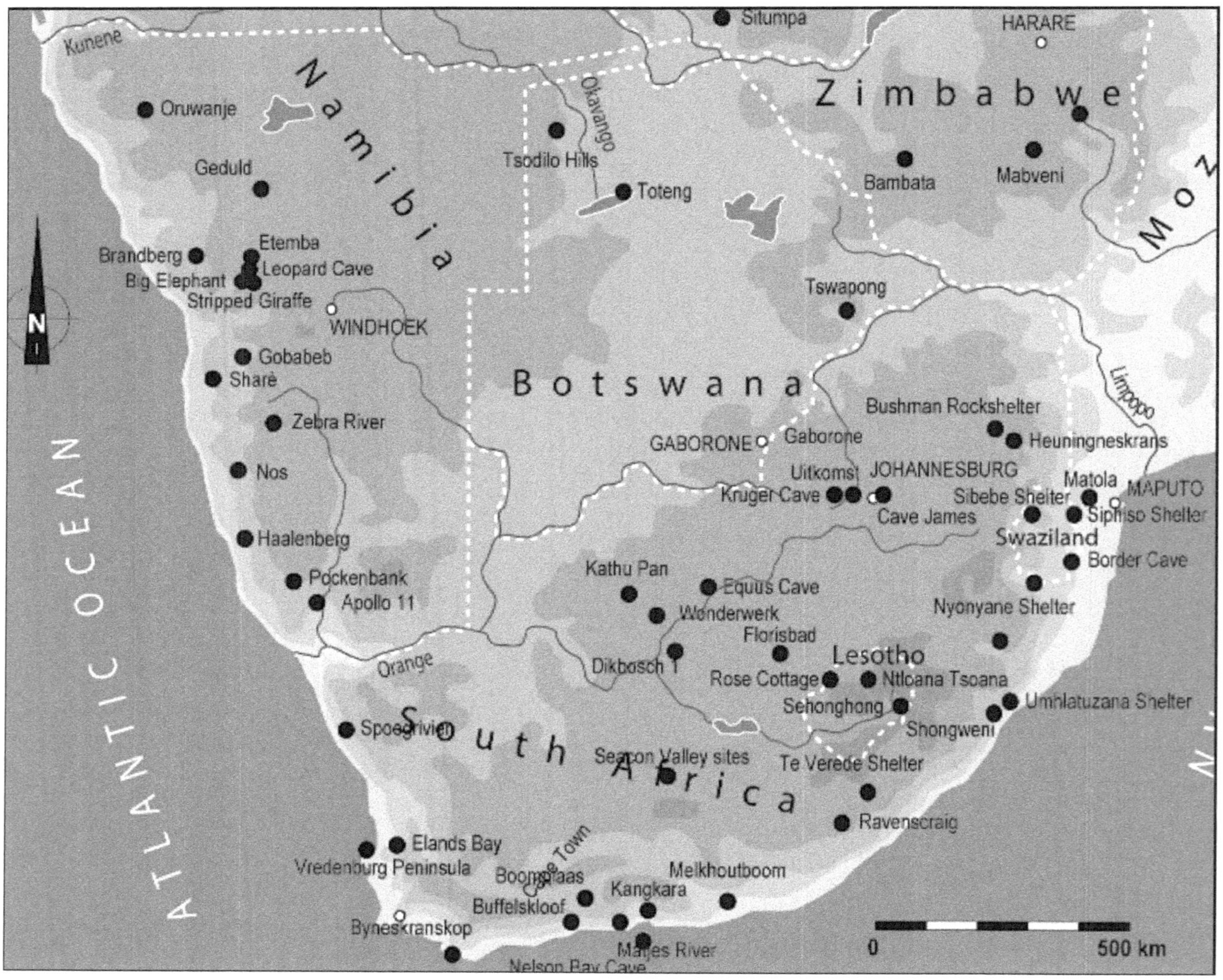

Figure 6. Map showing the density of main LSA sites in Southern Africa (after Pleurdeau et al. 2012).

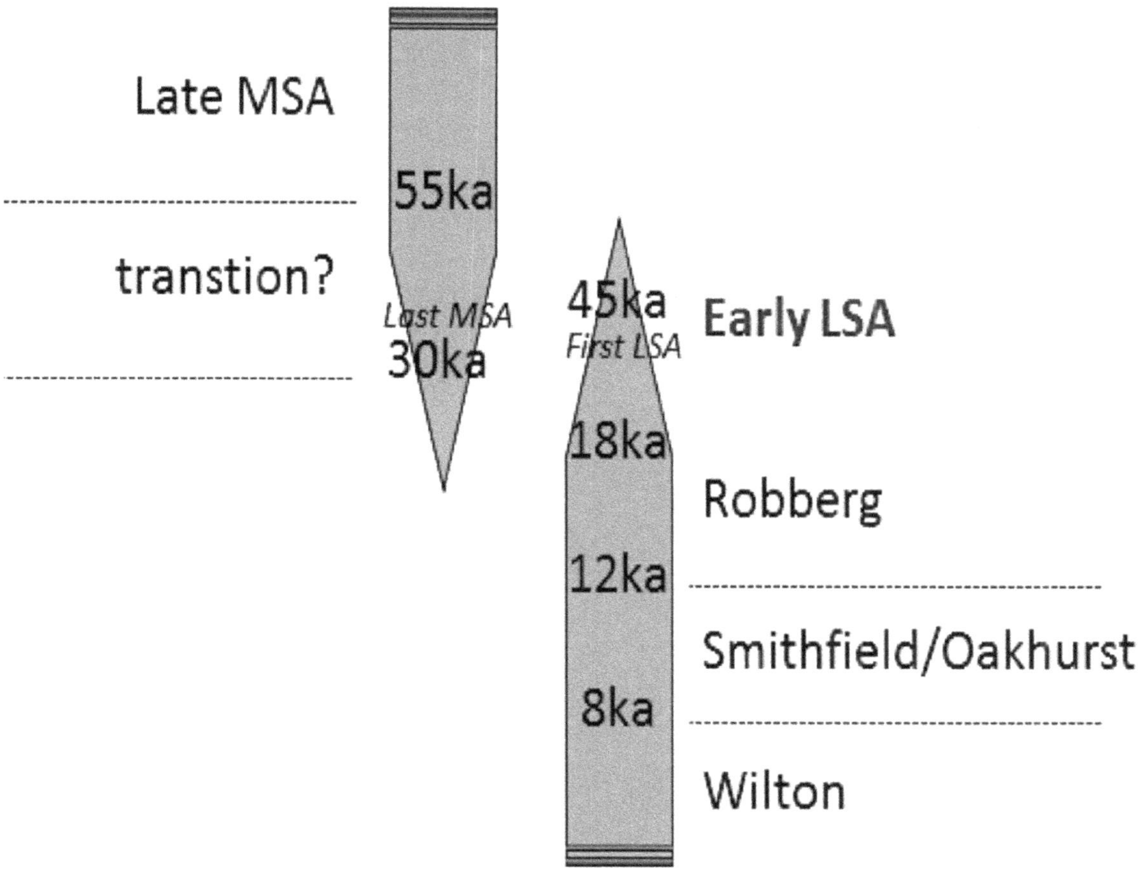

Figure 7. Synthetic cultural chronology of Late MSA and LSA traditions in Southern Africa. Photo: D. Pleurdeau.

1929) the Smithfield and the Wilton industries were often selected as stratigraphically uncontrolled contexts (Mitchell 1997; Mitchell 2002). The Wilton industry was named after the two rock shelters in the Albany district of the Eastern Cape where it was first identified. The Smithfield industry was suggested to be an autochthonous development determined by raw materials and unspecified ecological factors (Deacon, 2017). The challenge arising is fitting human populations and their vast cultural contexts into such industries. Interpreting stone tools requires discerning changes in subsistence strategies and settlement patterns during successive epochs through Pleistocene-Holocene, moreso that the Middle Stone Age (MSA) continued for at least 25 000 years after the appearance of the LSA. The Smithfield conforms to open-air sites while Wilton was found in rock shelters in the Cape and around the Limpopo. The differences between the two were initially explained in terms of geographical location and access to raw materials before the conception of functionality and temporal differences were added. But of recent, with a much clearer picture on Southern African palaeoenvironments, LSA studies have increasingly interpreted the Late Stone Age period beyond specific site and material culture alone and considered changes and significant regional or local variations.

The Robberg industry (See MIS2 in Table 5) was the last Pleistocene technological expression of the LSA and dates to 22,000/21,000 to 12,000 BP. The lithic assemblages are distinct for their abundant microlithic backed tools, bladelets, single platform bladelet cores, unifacial scraper industries, bipolar cores, backed microliths and other formal tools (Deacon 1984, Mitchell 1988, Deacon1984, Thornton Barnett 2013). Also found in association with these lithics are ostrich eggshell beads, bone points, bored stones and worked tortoise shells. Most if not all of the assemblages come solely from caves or rock shelters (Mitchell 1998). The identification and inclusion of surface sites is indispensable for understanding late Pleistocene settlement-subsistence systems, but is complicated by the fact that, the LSA assemblages are distinguished by the variation in artefact and raw material frequencies and artefact size, rather than the presence/absence of specific type-fossils, (Mitchell 1998).

The Holocene Classic Wilton Industry, on the other hand, is a fully-fledged microlithic industry ranging between ca. 8000 and 3200 BP. It is dominated by apt tools like backed microliths and small convex scrapers (Figure 8), occurring in association with bone tools and wood and shell artefacts. This period is followed by the post-Classic Wilton emerging in the Late Holocene age dating to between *c.* 2700 and 100 BP. Post-Classic Wilton can be divided into a pre-pottery stage predating *c.* 1800 BP and a pottery phase that dates from *c.* 2000 BP onwards. The post-Classic Wilton industry has regional variations

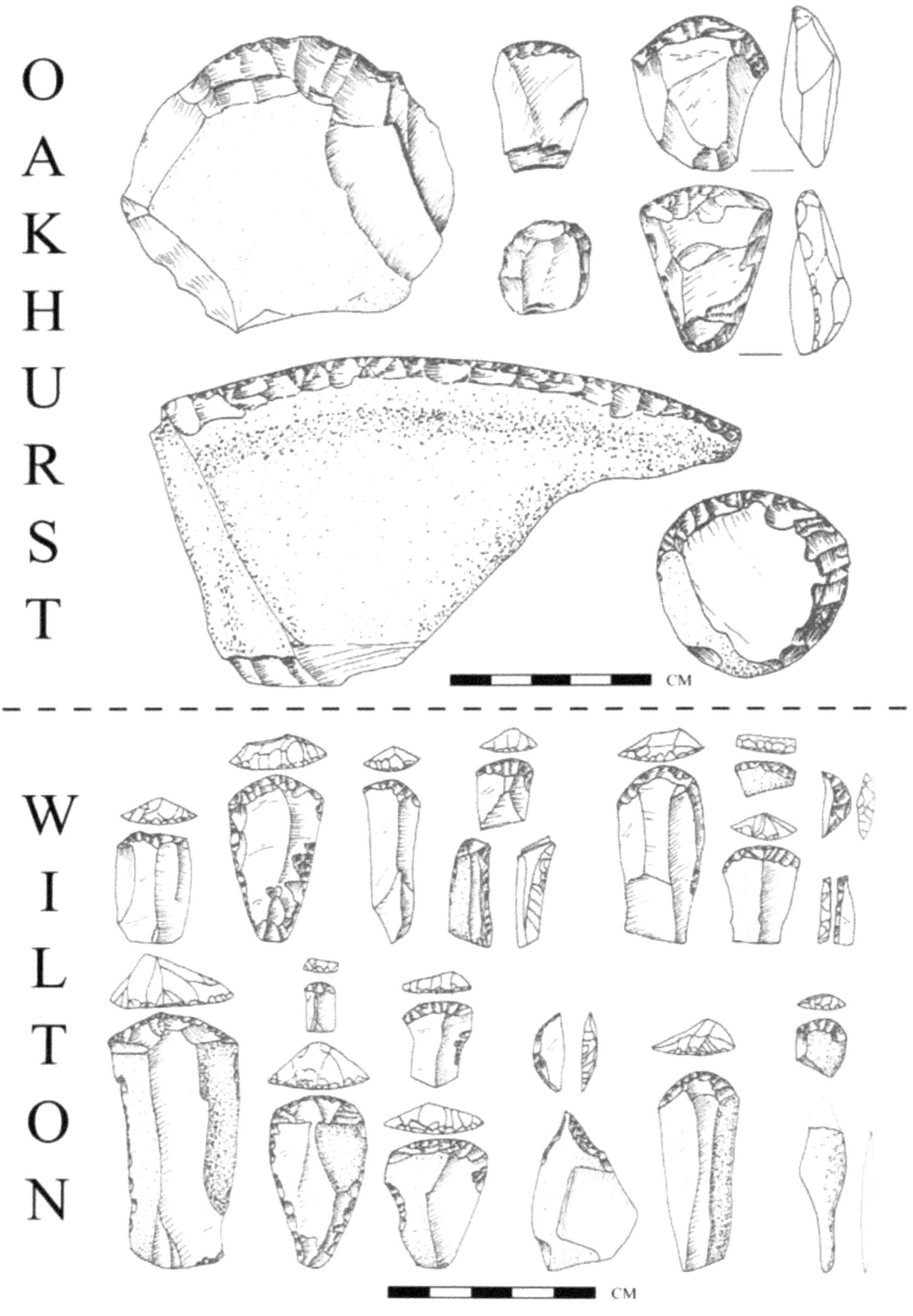

Figure 8. Variances between Oakhurst and Wilton artefacts from southern Africa. (Modified, after Forssmann 2019, depictions based on Deacon 1984).

and is more diverse than the Classic Wilton. In various areas, it is still dominated largely by a microlithic industry. However, in some other regions, it is associated with long end scrapers rather than with microliths (Deacon 1984). Post-Classic Wilton assemblages have among others been identified from the sites of Rose Cottage Cave and Tloutle in Lesotho.

The Oakhurst Complex dates to between 12,000–8000 cal. BP in most of Southern Africa and corresponds to the end of MIS2 and the early portion of MIS1 (Deacon 1984, Deacon and Deacon 1999, Mitchell 1997, 2002). The lithic assemblage is distinguishable by the presence of large scraping tools, with microliths absent (Deacon 1978, 1984, Mitchell 2002, Sampson 1974). Several regional variants exist including Lockshoek in the Eastern Karoo near Blydefontein, Kuruman industry in the Northwest Cape (Humphreys and Thackery 1983) and Coastal Oakhurst (Deacon 1976). The assemblages of the latter two variants contain abundant bone tools, possibly used as projectile tips (Wadley 1989).

In the past few decades, the definition of the Late Stone Age period has seen a shift from just emphasis on ecological adaptations, population mobility and systematic palaeoenvironmental change. Emerging in its place is more socially dynamic concept including social organisation and ideologies that go beyond just tool crafting and use. Still, the transition between the MSA and the LSA technologies of Southern Africa remain relatively sketchy (Mitchell 2002). A curious example is the seemingly present link between historical practices of the San communities of the Kgalagadi to Holocene hunter-gatherer LSA assemblages.

2.2. The Holocene LSA and its place among the history of the peopling of Southern Africa

From around 4000 BP a demographic rise is reflected in certain areas by the increase of the number of sites, the production of ceramics and the presence of domestic animals from ca. 2000 BP.

Some exceptionally preserved organic items that characterise more recent LSA as well as San material cultures, including bone points and digging sticks, were already in use at Border Cave around 44 000 BP (d'Errico & Backwell, 2016).

Theories that can confidently piece together the early development and evolution that constitute the Late Stone Age cultural and transitional dimensions and that include people remain illusive. Until recent decades this could be attributed to the lack of extensive ethnographic work and rock art research. Today, nevertheless, the LSA is still built on materialistic-based research in different parts of Southern Africa. For example, research has yet to arrive at a mutual conclusion, much less an approach, in understanding and contextualizing the so-called "herders"; a fact that makes it a subject worthy of curiosity. LSA research scholars, however, have devised a number of models around migration and exchange or assimilation of livestock, goods and ideas.

The migration model accentuates that proto-Khoekhoe communities who are the possible ancestors of the Khoe-speaking herders were responsible for the introduction of new cultural and economic subsistence strategies in territories occupied by Khoisan hunter-gatherers (Pleurdeau et al. 2012, Sadr 2008).

Numerous questions remain unaddressed. Significant climatic and environmental changes took place during the terminal and post-Pleistocene periods concerning numerous cultural developments in several parts of Africa and the world. Whereas elsewhere more sedentary organised lifestyles and developments in food productions were the case, in southern Africa, such practices arrived much later. The absence of food production traditions including the adoption of new cultivated crops and the rearing of domestic animals leave unresolved the identities of the occupants of the southern part of Africa and their socio-cultural and socio-ecological practices during the late Pleistocene (Deacon, 2017).

The current debate on the early peopling and evolution of Southern Africa including during the Pleistocene and the beginning of the Holocene is hampered by a lack of evidence. The most notable among them is the absence of domestic stock and anatomically modern human fossils in these earlier periods (Pleurdeau et al. 2012). Meanwhile, in Southern Africa, the onset of the Holocene brought better climatic conditions and the diversity of plant and animal species increased (Plug, 1997). The Pleistocene/Holocene sites around the Drakensberg and Lesotho show an increase in faunal species abundance which is attributed to the improved climatic conditions that follow the period after the Late Pleistocene. The topography of some of the sites could have played a major role in species distribution. For example, from the results from the eastern highlands of South Africa and Lesotho, it could be hypothesized that the sites at higher altitudes (Plug, 1997) might have been unsupportive for certain species because a colder climate shortened the summer and limited grazing areas. This pattern, which might have been ubiquitous in the Late Pleistocene, can still be observed today with some animals such as eland species being forced out of their territory during the winter.

2.3. Chronology and the place of herders: the advent of a Southern African 'Neolithic'?

The last 2000 years suggest a coexistence of different subsistence strategies; based on the introduction of herding practices and/or the continuation of hunting and gathering practices. Even though by now there is an agreement between different disciplines (linguists, anthropologists, archaeologists) that the sheep reached Southern Africa around 2000 years ago (Sadr et al. 2017; Pleurdeau et al. 2012; Robbins et al. 2008)(Figure 9) what is still unclear is the origin of these animals as well as a consensus on the identity of the "herders" who kept them.

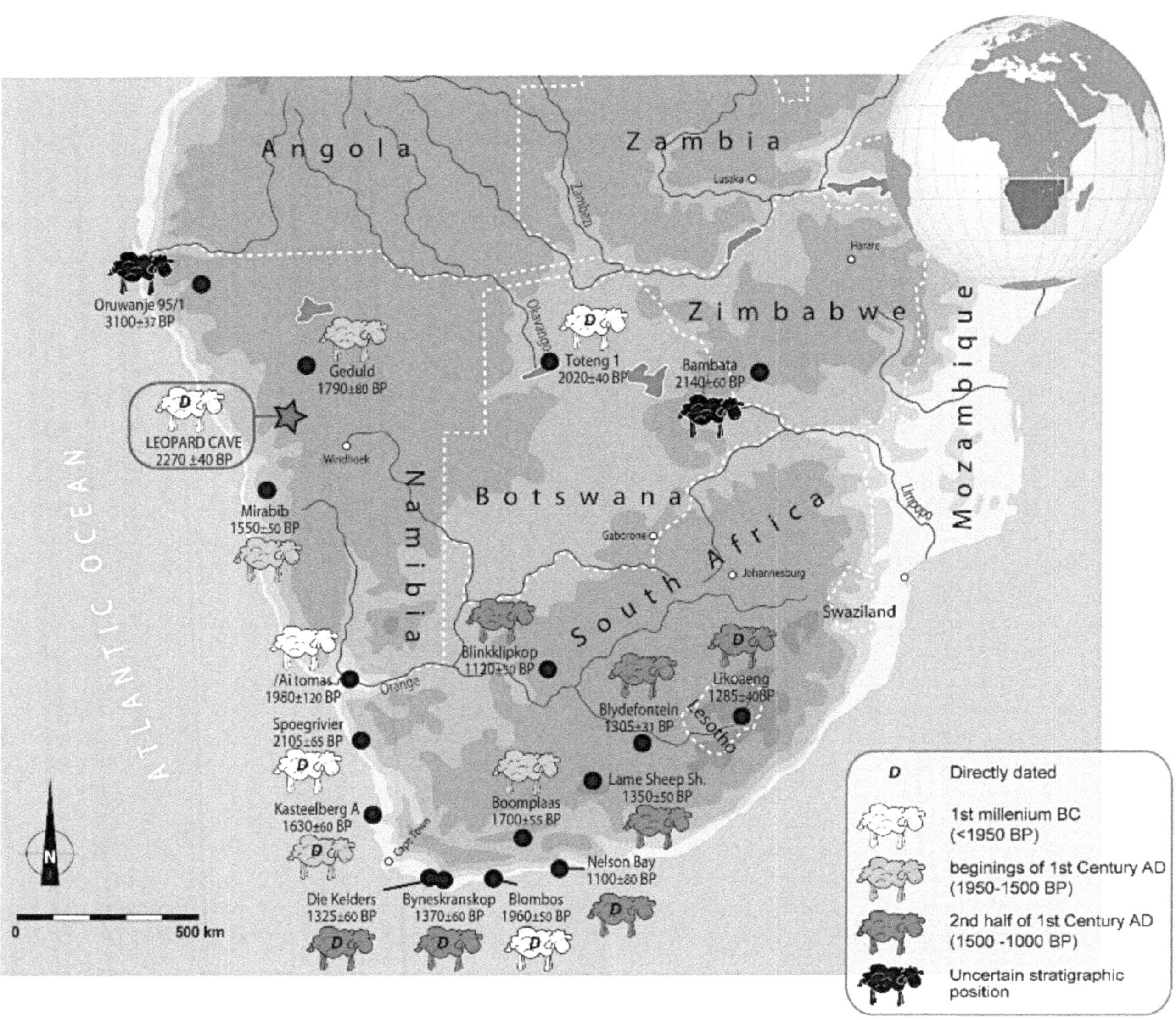

Figure 9. Later Stone Age sites of Southern Africa with early evidence of caprines (after Pleurdeau et al., 2012).

In southern Africa, as is the case in other parts of Africa, archaeological sites occupied by pastoralists are determined by a prevalence of stock and other mammalian bones (Sadr, 2013) . The earliest evidence from north-western Botswana puts them two to three centuries before the well documented iron-using Bantu-speakers' arrival from Eastern Africa into southern Africa, suggests that the Bantu-speakers are not responsible for the introduction of cattle in LSA southern Africa and its economies at the time.

In 1979, sheep remains from Die Kelders in the Western Cape dating back to c. 2000 years ago in association with pottery were discovered, suggesting the possible contemporary nature of the earliest sheep and ceramics (A. B. Smith, 2013). Since then, many more sites on the west and south coasts have produced dates for sheep between 2100 and 1860 BP. Thus the most significant aspect of the LSA period is the presence of archaeological sites with evidence of domestic stock and other mammalian bones (Sadr, 2008; Sadr et al., 2003; Sadr, 2013). Ultimately in the 1980s, "pastoralists" sites with an abundance of sheep bones, stone enclosures and ceramics believed to belong to the Khoekhoe were found at Kasteelberg in the Western Cape and in the Karoo region of South Africa. The absence of domesticated livestock prior to this period suggests possible migrations and introduction of animals southwards probably accompanied by human migration (Orton et al., 2013; Pleurdeau et al., 2012; Sadr, 2008; Sadr 2013). But at the same time, their presence in the region does not necessarily mean that they were accompanied by human migrations as that cannot be demonstrated. Interestingly while sheep do not appear where they might be "expected" (i.e in open air animal enclosures, inhabited by the Khoekhoe), evidence of early sheep have been found in rock shelter sites with a material culture that correlates with San hunter-gatherers. Some theories attribute the origins of herding to the influx of genetically distinct herders into the subcontinent (Figure 10) while other schools hold the notion that herding stemmed from the acculturation of the local hunter-gatherer communities through small clusters of migrant

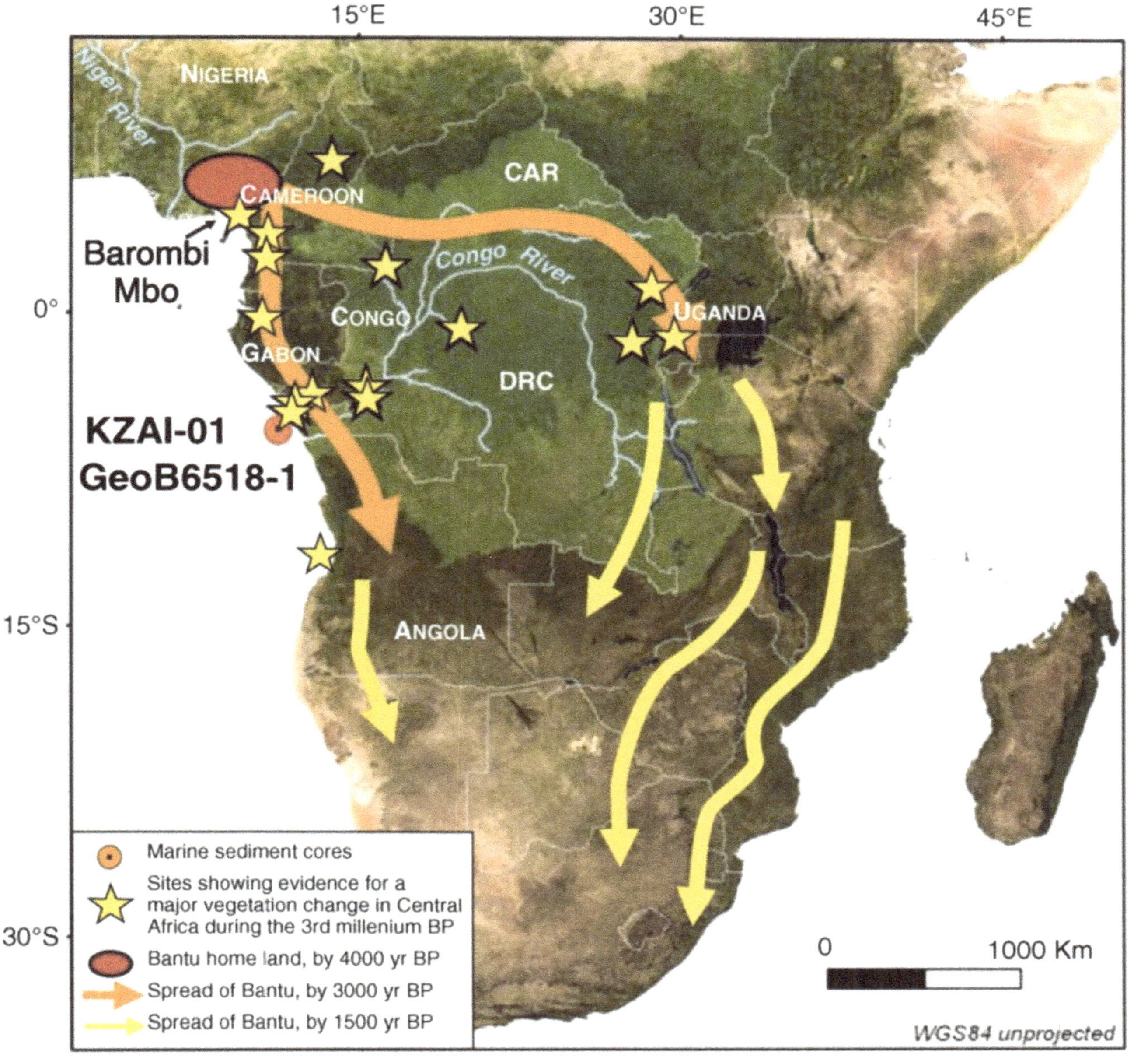

Figure 10. One of the proposed migration models showing routes from the west into Southern Africa possibly with livestock. Image source: Abagond.wordpress.com, 2018.

herders who were genetically absorbed into the gene pool of the already established populations (Sadr 1998; Stynder 2009; Pleurdeau et al. 2012). Nonetheless, a number of factors can play a role in these migrations; the underlying driving factors being social and environmental (B. Heine and König, 2008; Mitchell, 2010) for both animals and herders (Pleurdeau et al., 2012).

This, therefore, puts into question the idea of the Khoekhoe speakers as the introducers of the livestock through a migration into the Cape. At the same time, researchers perhaps should stop searching for what is absent, as the absence of evidence might mean the evidence of absence. "Pastoralists" sites might never materialise because nomadic pastoralists were likely to inhabit temporary locations and that the few sheep remains found in San rock shelters were likely acquired from the nearby Khoekhoe pastoralists through theft or exchange. Perhaps the San, too, may have later become herders (Sadr, 2008; Sadr 2013). Hypothetical Proto-Khoe populations of the "pastoralists" (Khoe speaking) encountered by the first Europeans who arrived on the southernmost coast of Africa from the end of the 15th century onwards is often thought to be the potential herders. Several recent genetic and linguistic studies support this hypothesis, although the place of origin of these early herders and/or stock is still widely debated.

Evidence of domesticated animals (mostly of sheep and very few cattle) with reliable AMS radiocarbon dates prior to 1500 BP are very rare. Currently, the most reliable direct dates of domesticated animal presences are from Spoegrivier (2105±65 BP) Blombos (1960±50 BP), and Kasteelberg (1630±60 BP) in South Africa and easily the oldest in Southern Africa, the site of Toteng in north-western Botswana (2020±40 and 2070±40 BP). The remainder of the sites provides morphologically identified caprines with dates assigned from stratigraphic associations, supposedly from LSA contexts but without direct dates as is the case at Geduld and Mirabib in Namibia dating to between 1950 and 1500 BP, Boomplaas, in South Africa and other later sites such as Lame sheep and Blinkklipkop in South Africa falling between 1500 and 1000BP (Pleurdeau et al. 2012, Smith et al. 1995). By applying biochemical proteomic analysis to faunal remains from Leopard Cave (Le Meillour et al., 2018) show that bone remains first thought to have belonged to domesticated caprines (sheep/goat) in reality correspond to wild antelopes (see below: Leopard Cave). This discovery, made possible through the use of innovative techniques, raises the question of the identification of caprine remains on other sites and thus of the herder versus hunter model.

Imminent questions include:

a. What constitutes the Holocene Late Stone Age period, in terms of subsistence and hunting strategies, networks or material culture, particularly in the context of the arrival of possible herding practices?
b. is the apparent scarcity of early/middle Holocene LSA occupation a reflection of reality, or a preference for hard-to-find-and-date open-air sites (the not so preserved type of sites) or rather an absence of herding practices at this time?

If recent research projects have indeed been successful in merging different disciplines (archaeology, linguistics, genetics) with a theoretical framework and address certain questions, the global picture of the LSA is yet to be understood (Smith 1993). Thus, the research doors on this subject remain open.

A few hypotheses try to explain the first occurrence of herders and herding practices in Southern Africa:

a. Migration from East Africa by pastoralists who arrived via Northern Botswana with livestock and either mixed with the original populations of western southern Africa (genetic mix) or settled in the region without interacting with local hunter-gatherer groups.
b. The introduction of livestock by exchange with herders or by the spontaneous wandering of cattle from other regions (Henn et al. 2008, Mitchel et al. 2010, Sadr 2013).

If the question is "did the Bantu speaking agro-pastoralists bring both sheep and cattle into southern Africa?" then the second hypothesis would prove insufficient to answer the question. Firstly, there is no archaeological evidence to support this assertion and secondly, livestock without guidance could not have wandered that far. In any case, the debate remains even though archaeological, linguistic and genetic data support the former hypothesis (demic diffusion e.g. the arrival of new populations introducing their livestock and herding practices).

The small quantities of faunal remains corresponding to domesticated animals at sites dating to around 1000 AD are somewhat problematic when defining these sites as pastoralist. Sadr (2013) even suggests that certain practices resulted in the elimination of animal bones from the sites. Another possibility is that only a very limited number of cattle were kept. The consumption of livestock became elevated 2000 years ago but the lack of both faunal representation and human remains for isotope studies makes it difficult to prove this hypothesis.

2.3.1. Subsistence strategies

Food security during the LSA was heavily dependent on the environment. The Karoo and Cape biomes are rich in carbohydrates-rich geophytes, and fruit trees prevail in the Savannah biomes, while melons and tubers are prevalent in the Kgalagadi area (Clark & Walton, 1962). These geographical differences of biomes might account for the presence of certain remains of plant or faunal markers in some areas but not in others. Plant remains are more prevalent in the Mid-Holocene rock shelters dominated by the family Iridaceae in the Cape area. Even though the Holocene lacks a detailed record on the use of marine and littoral resources, MSA sites like Klassies River and Eland Cave (Reynard & Wurz, 2020; Brenner et al., 2020) show the use of these resources during the Pleistocene until the Terminal Pleistocene possibly with a Pleistocene/Holocene transition. Holocene site distribution for subsistence information derivation is very important to take into consideration. The period around 6000 BP has a very low number of sites, namely in Lesotho and the Transvaal that correspond to the Mid-Holocene aridity, a factor that might have inhibited human settlements at the time (Mitchell, 1997; Fitchett et al., 2017).

Given that the amount of faunal remains in the Pleistocene surpasses that of the Holocene, Mitchell (2002) suggests that Pleistocene inhabitants consumed more meat than their Holocene counterparts, who were less reliant on meat-based diets. Even though evidence shows that both were foraging, the late Pleistocene populations mostly hunted large mammals like equines and eland. Holocene populations appear to have been less reliant on foraging than previously thought with the discovery of large grazer remains such as wildebeest and hartebeest in their assemblages. Domestic cattle are also present in Namaqualand in South Africa suggesting that cattle and sheep may have arrived together in the Cape region more or less 2000 years ago. Deacon (2017) speculates that wider investigations further to the south might show that cattle were also kept on contemporary Namibian sites.

2.4. Leopard Cave (Erongo, Namibia)

Leopard Cave (S 21°34′22″, E15°33′18″) is located on the farm of Omandumba West, in the Omaruru region, northwestern Erongo Mountain, ~200 km northwest of Windhoek (Figure 11). The site is located ~ 5 km from the farm, and 1.7 km south of the Fackelträger site, excavated by Dr Wendt in the 1960s (Wendt, 1972). The site is a rock shelter measuring c. 60 square meters (about 8 meters wide) and opens at an altitude of 1256 meters in a Cretaceous granite formation. The shelter opens to the south, facing the high peaks of the Erongo.

The site was discovered in 2006 in a Franco-Namibian cooperative project. From 2007, several field seasons have permitted the excavation of a surface of more than 8 m^2 within the shelter (Figure 12) and almost 4 m^2 outside the shelter. The unearthing of a 2 m deep archaeological sequence has allowed the discovery of a succession of Holocene human occupational levels (Figure 13).

To a large extent, this sequence is dated to 5000–2000 BP. It allows for the documentation of hunter-gatherer occupations from the Later Stone Age, particularly based on an opportunistic production of stone tools, bone weapons, and of non-lithic and non-utilitarian productions such as beads or pendants. The excavations also provided numerous artefacts linked to pigment processing, such as crushing and grinding tools as well as concentrations of pigment blocks. Physico-chemical studies, including pXRF analysis (Mauran et al., 2019), have shown the link between the pigment used for parietal representations on the East and North walls of the shelter and the residues on archaeological tools sourced around the area.

2.4.1. Stratigraphy and dating

The latest excavation seasons (from 2014–2019) have allowed for a broader comprehension of the chrono-stratigraphic sequence of Leopard Cave (an E-W section of more than 3 m in height). The sedimentary sequence comprises three stratigraphic units, including one which seems to correspond to an accumulation of granite blocks without any visible anthropogenic remains. The thick upper stratigraphic unit (z=150–250 cm) is composed of a succession of horizontal powder-silted layers born from the granite boulders break-up mixed with anthropogenic ashes. It corresponds to a succession of recurrent

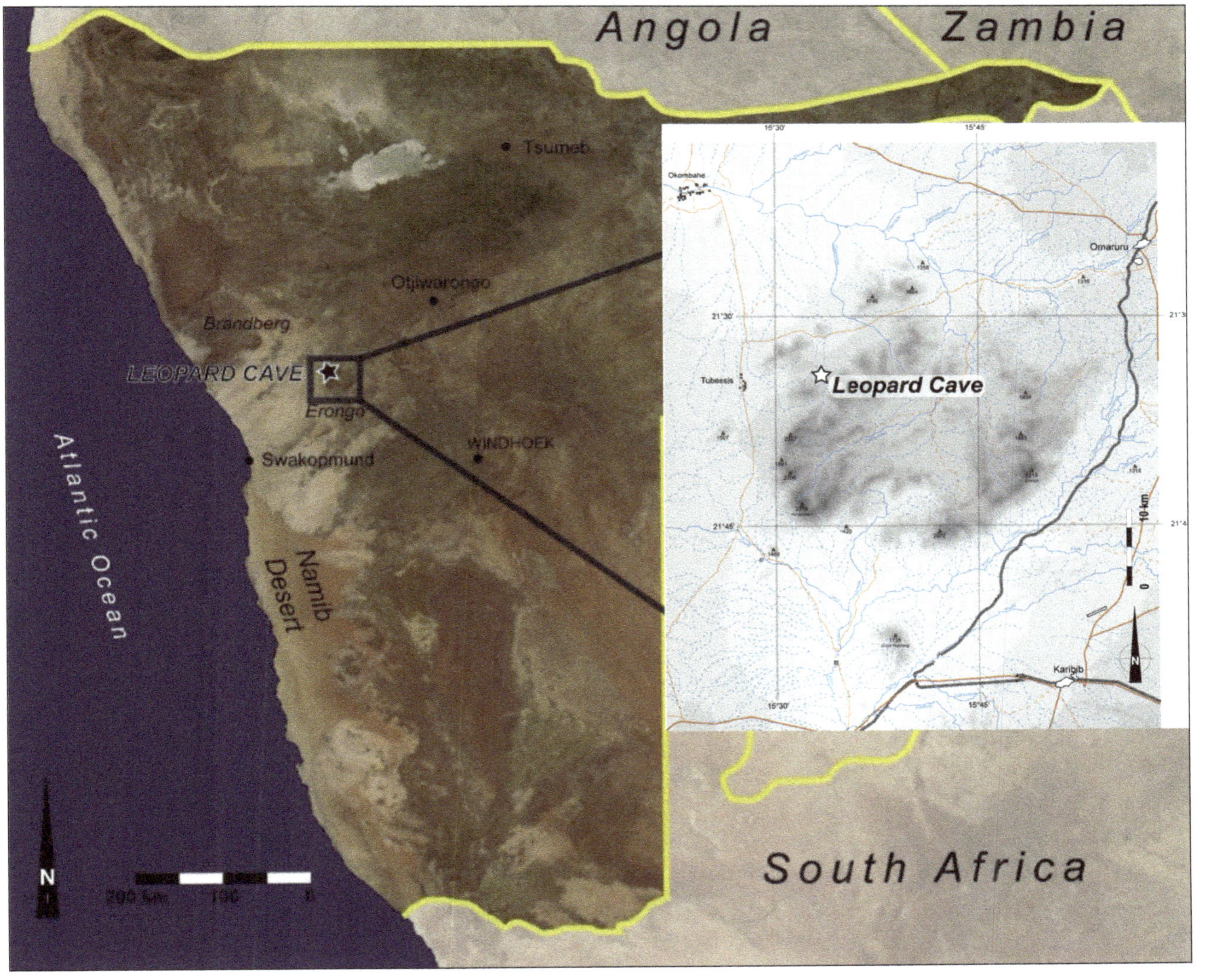

Figure 11. Location of the site of Leopard Cave (Namibia) (D. Pleurdeau).

Figure 12. View of Leopard cave in 2012 (l) and location of the excavations areas (r).

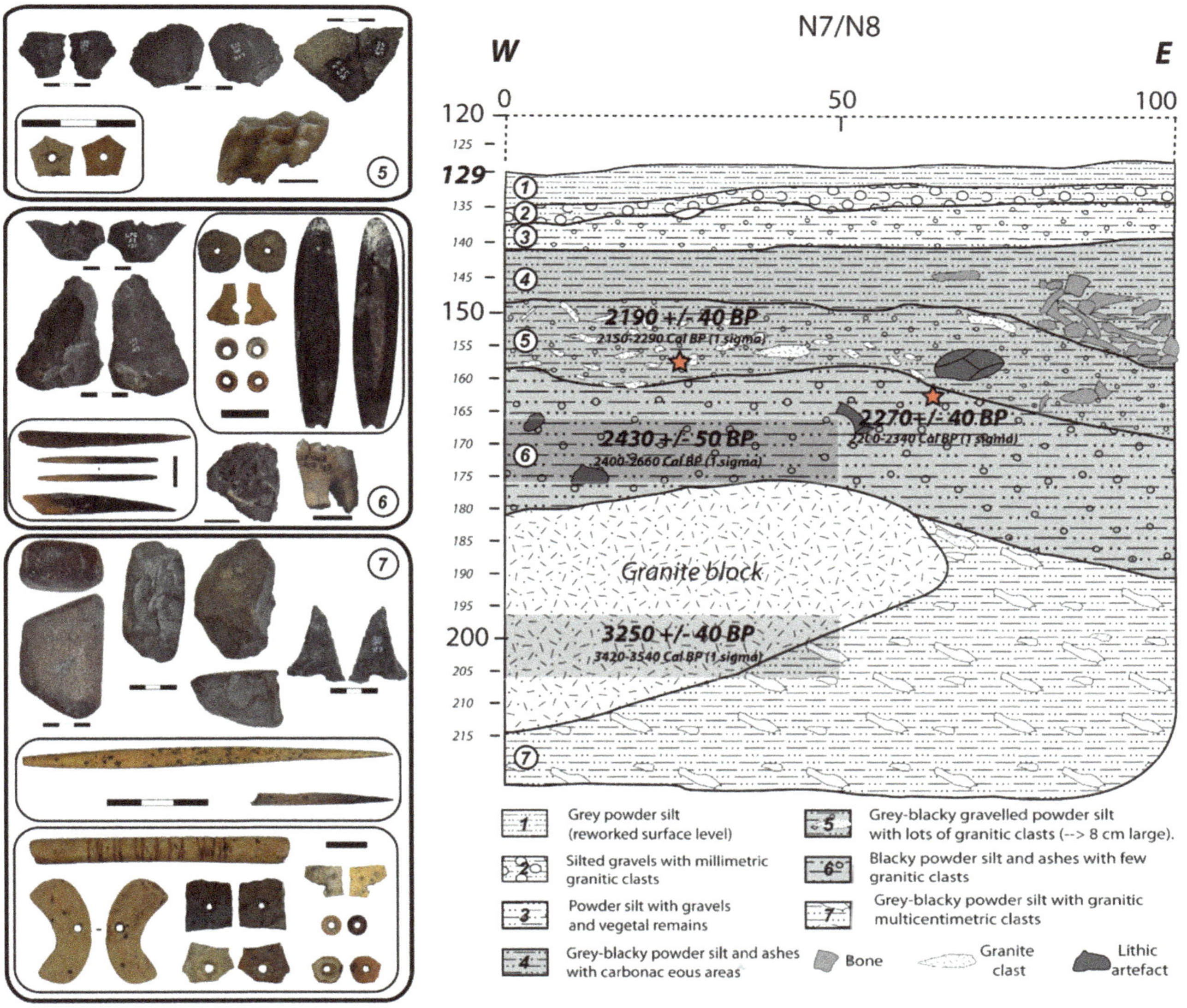

Figure 13. East-West transversal section and an example of recovered material from Leopard cave. (after Pleurdeau et al. 2012).

occupation layers. In addition, some levels present numerous and thick clastic granite residues. Finally, in the lower half of the sequence, a level of human occupation indicators is still visible at the base of the sequence more than 1.8 m below the level of the current sediment surface. The last sedimentary layers reached in 2017 present a different structure from the overlying levels, with sediment indurations and the absence of the ashy matrix found in the higher levels. A very different lithic assemblage, with technical characteristics similar to those of the MSA, was discovered at the lowest levels.

The stratigraphic sequence, divided into three main complexes (Figure 14), could be summarized as follows:

- Complex I (z=150[7]–250 cm): This complex, which comprises mainly of silt and ashes of anthropic origin and abundant vegetal remains and charcoal, is divided into nine sub-horizontal mix layers (now labelled from a (top) to j (bottom)). The presence of ash/charcoal and a large number of artefacts recovered from this stratigraphic complex demonstrate the existence of intense anthropic activities.
- Complex II, between z= –250 cm and z= –280/300 cm, is associated with granite block collapse episodes. The few artefacts discovered within these layers could correspond to upper layer infiltrations and infilling between the blocks and not to in situ human occupations layers.
- Complex III (from z=280/300 cm) is made of an indurated arenitic matrix, without the presence of ash or organic material. The substratum has not been reached.

The radiocarbon ages (Table 6) obtained since the beginning of the excavation of the site have allowed us to establish a chronology of human occupation, ranging from ~3700 cal BP (z = 250/260) to ~2000 BP cal BP (z = 150/160) for the upper part of the sequence. Two main occupational phases within the upper Complex are highlighted by the

[7] The altimetric level 0 is located ~150 above the nowadays sediment surface

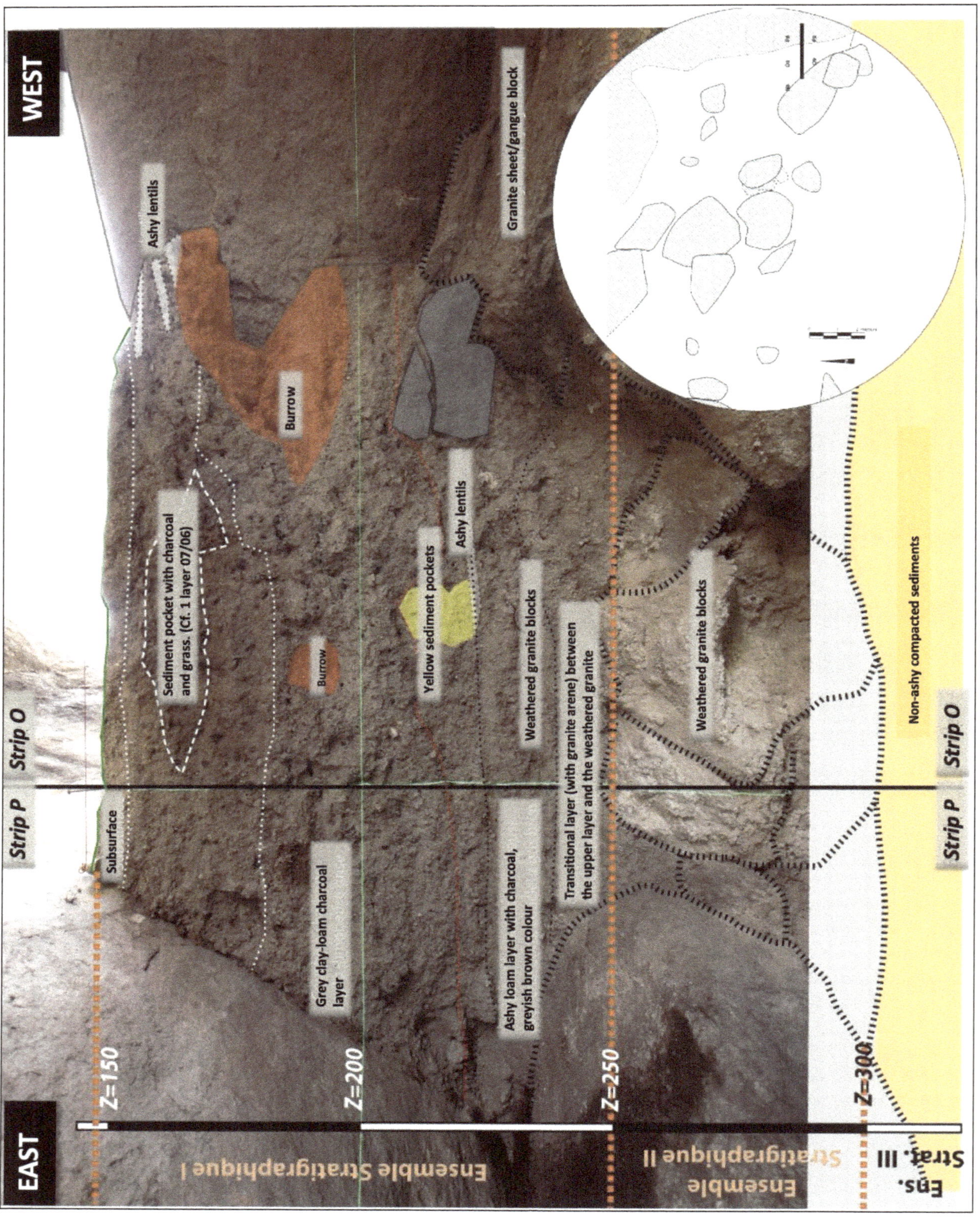

Figure 14. Stratigraphic section of 8/9 transversal bands (Squares NOP8/NOP9) Image credit: D. Pleurdeau 2012, modified).

Table 6. Radiocarbon ages of the archaeological sequence of Leopard Cave (after Pleurdeau et al. 2017)

Lab#	Year	Square		Depth (cm)	Nature	Conventional Radiocarbon Age BP	Age Cal BP (2 sigma)
SacA34262	2013	N6	Unit I	143,5	Charcoal	2150 ± 30	2160 ±145
UBA-38617	2018	P7	Unit I	150–160	Seed	1991 ± 33	
Beta – 270163	2010	N7a	Unit I	157,5	Tooth	2190 ± 40	2165 ±195
Beta – 270164	2010	N7b	Unit I	163	Tooth	2270 ± 40	2267 ± 96
UBA-30945	2016	P7	Unit I	175,5*	Charbon	2209 ± 33	2187 ± 125
Beta – 236963	2009	N7a	Unit I	166–176	Charcoal	2430± 50	2515 ±185
SacA34261	2013	M6	Unit I	159	Charcoal	2565 ± 30	2620±125
CA_DAT_AD63C	2017	P8	Unit I	168	Charcoal	2105 ± 45	2100 ± 185
UBA-30946	2016	P7	Unit I	192	Charcoal	3064 ± 32	3212 ± 135
Beta – 236964	2009	N7a	Unit I	196–206	Charcoal	3250 ± 40	3460 ± 100
CA_DAT_AD63A	2017	P8	Unit I	203**	Charcoal	2267 ± 45	2230 ± 115
Beta – 236966	2009	N7	Unit I	254–279	Charcoal	± 40	3342 ± 105
SacA42295	2015	N7	Unit I	263,5	Charcoal	3465 ± 30	3700 ± 125
SacA42299	2015	N7	Unit I	271,5	Charcoal	3000 ± 30	3110 ± 120
CA-DAT_AD63B	2017	P8	Unit II	240–250	Charcoal	4508± 45	5090± 210
SacA53511	2018	O7	Unit II	290	Charcoal	4945± 30	5650± 60
SacA513109	2017	O7	Unit III	317	Charcoal	13040 ± 70	15540± 270

C14 dating sets, as presented in Figure 15, one between c. 3800 and 3000 cal BP and a second from c. 2700 to 2000 cal BP. Only one age is currently chronologically adjoining the third unit (from ~300cm). This age, as well as the (few) lithic remains recovered (N=52), which have different technical features (elongated products, bifacial points…) from the upper unit, indicate a terminal Pleistocene period of occupation. Ongoing excavation of this lower unit, towards the substratum, will probably emphasise this development towards a possible MSA-LSA transition.

2.4.2. Lithic materials

The lithic industry represents more than 1,000 plotted artifacts from the upper unit. With the notable exception of a few pieces of allochthonous chert, it comes from the exploitation of local quartz nodules (~ 30%) and basalt (~ 70%). The thermic actions, resulting from the intensive presence of hearths over a large part of the sequence (ashy matrix of the sediment) affected the original nature of the remains of a large part of the material (very often cracked or split). These alterations often make the technological reading of the pieces difficult. The first general study had shown that the majority of quartz pieces are fragments and/or debris, resulting from intensive debitage. The nuclei are dimensionally very small (a few centimetres) and do not have a diagnostic organization of the debitage. The fragments, as well as these nuclei, show above all the use of a bipolar percussion on the anvil, having caused a major fragmentation of the material. The basalt pieces, which are often altered (thermal actions during or after burial), provide more precise information that the operative patterns involved several reduction modalities: Flaking debitage from unipolar reduction with opposite edges; elongated flakes debitage from convergent reduction; and heavy-duty tools profusion (with sharp and/or pointed edges) (Guillemard 2018. In. Pleurdeau et al.)(Figure 16).

2.4.3. Faunal remains

The excavations also yielded very abundant faunal remains (over 6500). In general, the remains are poorly preserved, with more than 50 percent that could not be identified due to the high fragmentation during the taphonomic processes (corrosion, bad weather and carbonization). These post-sedimentary processes might also have damaged the surface of the bones and thus limited the preservation of anthropogenic cut marks, only visible on 1 percent of the bones (Lesur in Pleurdeau et al. 2018).Fragments of the eggshell (mainly ostrich, but also other birds like guinea fowl) account for 66 percent of the identified remains. This high number of eggshell fragments shows the importance of this material to the inhabitants of the site. The presence of dozens of ostrich eggshell beads (OEB) illustrates their use for ornamentation. In addition, one-third of the eggshell fragments are burnt, potentially suggesting their consumption or use of the whole eggs in a cooking vessel. Among the vertebrate remains, there is a great

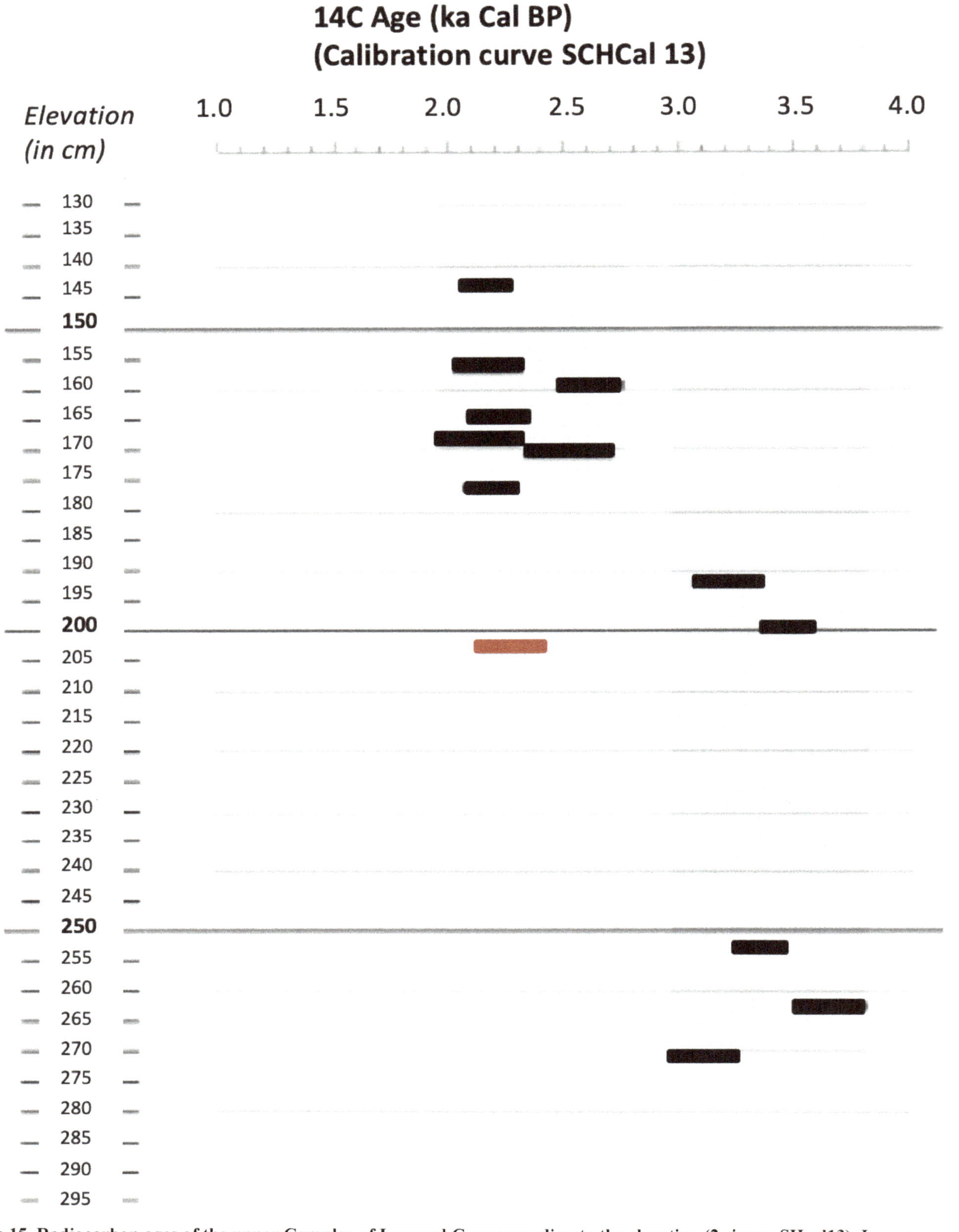

Figure 15. Radiocarbon ages of the upper Complex of Leopard Cave according to the elevation (2 sigma, SHcal13). Image credit: D. Pleurdeau.

diversity of taxa, especially among mammals. Of these, the majority correspond to medium-sized animals such as impala, springbok or caprines (sheep and goat). Smaller antelopes (cliffspinger and steenbok), warthog, hyrax, hare and small carnivores are also present. In addition to these, there was a discovery in 2012 of a pierced cat mandible, which cannot be identified as wild or domestic, and which shows no trace of possible consumption. The analysis seems to show that there is no major change in the list of fauna throughout the sequence. In addition,

Figure 16. Examples of lithic materials from Upper complex of Leopard Cave (after Guillemard 2018).

all identified species are still present in the area today, suggesting that environmental conditions were close to those of today. The presence of these species seems to attest that the environment surrounding the rock shelter consisted of three eco-zones composed of a soil covered with perennial grass, bushes and scrub, steppe trees as well as species of woody plants. The inhabitants of the site have exploited many available resources such as mammals but also reptiles and birds (Lesur in Pleurdeau et al. 2018). Due to the poor conservation of the remains, it is difficult to determine the techniques used by Leopard Cave occupants to exploit their environment. Medium-size animals such as impalas or springbok have been hunted, but no artifacts found suggest specific techniques. Some bones show signs of butchery but, again, their small numbers do not allow these activities to be modelled. Notably, the several bones morphologically attributed to domestic caprines (sheep *Ovis aries* or goat *Capra hircus*) found in the shelter, including three teeth directly dated around 2.3 ka BP amount to the first evidence of animal domestication in the entire Southern Africa (Pleurdeau et al., 2012, Figure 17). Like DNA, the peptide signature is specific to each species, or even to each individual. Proteomic analysis was therefore initiated on these dental remains, in order to discriminate the remains of Leopard Cave (goat or sheep), but also to develop an analysis protocol adapted to remains from arid environments (Le Meillour et al., 2018). The peptide identified as specific markers on the archaeological teeth of Leopard Cave is associated with *Antidorcas marsupialis* or springbok, a species of small wild bovid still present in these environments and already abundant on the site, indicating that it is probably necessary to revise the assignment previously made on morphological criteria. The presence of three markers specific to springbok indicates that the remains of Leopard Cave should rather be placed in the sub-family of antilopine rather than caprines (Le Meillour et al., 2018; Le Meillour et al., 2020).

2.4.4. Human remains

The first of the human remains discovered in the shelter is a human tooth unearthed in 2015, in layer 6 (z = 175–180). Radiocarbon dating of nearby charcoal has confirmed an age of approximately 2200 cal BP. This molar is in a poor state of preservation since only the dental crown, broken in 3 parts, has been found (the roots are totally missing). It is a second upper right decidual molar for which advanced occlusal wear and the presence of a well-developed interproximal contact facet on the distal surface would indicate an individual age of about 10–11 years. In the current state of discoveries made at Leopard Cave (i.e., no other immature human remains), it would be a deciduous molar naturally lost on the site. In 2018, a human zygomatic (Figure 18): has also been recovered from Square P8 (z=220–225), within the Upper Complex I.

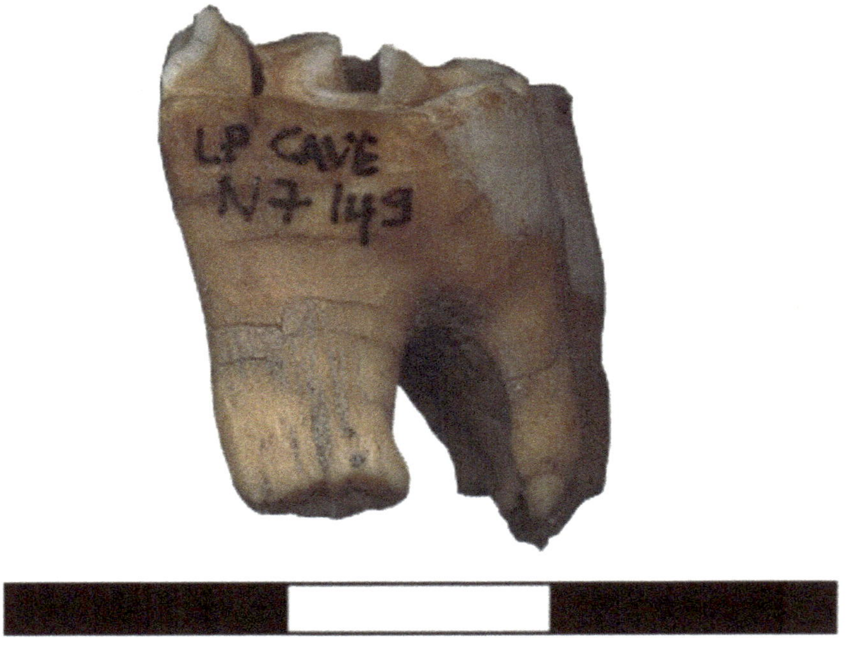

Figure 17. Small bovid tooth from Upper Complex I of Leopard Cave, initially identified as caprine. Photo credit : F. Detroit.

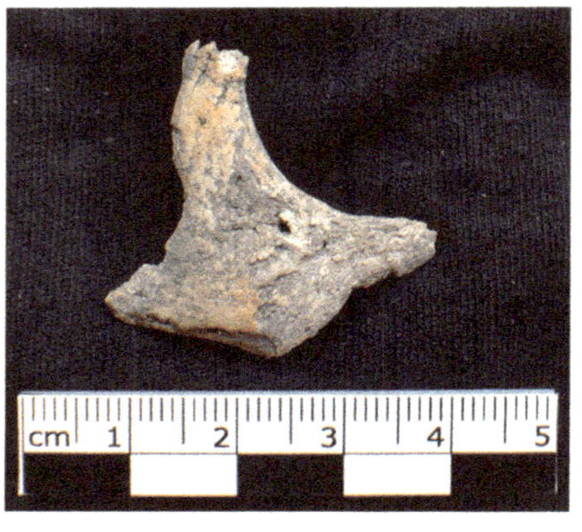

Figure 18. Human zygomatic from the Upper Complex I. Photo credit: F. Detroit.

Figure 19. Giraffe figure as depicted in the rock art. Photo credit: M. Mvimi.

2.4.5. Pigments and rock art

Leopard Cave is a rock-painting site, with several representations of animal figures (Figure 19: Giraffe figure as depicted in the rockart among other faded pictures on the wall). In the archaeological layers, hundreds of pigment blocks, as well as a large corpus of material associated with these coloring materials (e.g. passive and active grinding stones used for the transformation of the pigment blocks and suggesting an *in-situ* preparation of painting powder), and other objects carrying traces of pigments on their surfaces have been discovered within a large part of the sequence (between 3,5 and 2 ka cal BP) (Figure 20). The Erongo massif is home to thousands of ancient rock paintings most of which are concentrated in the granitic massifs of the Brandberg, Spitzkoppe, Twyfelfontein and the Erongo (Breuil 1955; Nankela 2017). A recent pXRF analysis completed by Mauran et al., (2019) has highlighted a high alteration of the paintings as shown by the presence of Sulfur and Calcium minerals in most of the analyses. They also suggest possible local sourcing of the raw material used in the paintings. The paintings are composed mainly of white and ochre and iron oxide red pigments, characterised by a presence of animals, and human figures associated with hunter-gatherer San populations of Southern Africa. Like many aspects of prehistoric research in Namibia, this area was neglected despite its rich rock art culture, possibly due to geological and topographic conditions that are less favourable to the

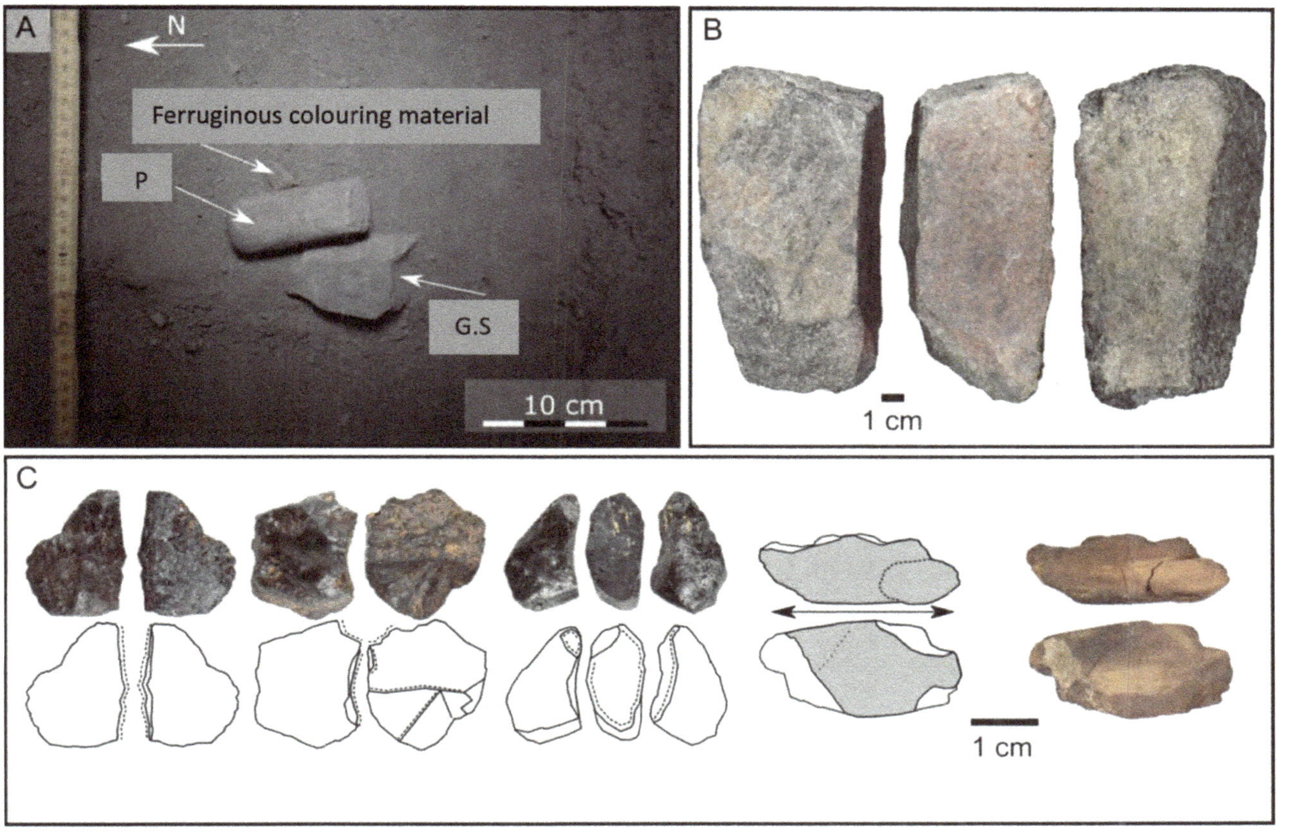

Figure 20. A) In situ pestle (P) and grinding stone (G.S) associated with ferruginous colouring material. B) Pestle covered in red residue and ferruginous colouring material showing evidence of use-wear (C). (after Mauran 2020).

Figure 21. Current environment in close proximity to the site of Leopard Cave. Photo credit: M. Mvimi.

accumulation of sediment and the registration of the stages of site use. However, Leopard Cave alone among rock art sites in the region yields painting raw materials as well as tools used for the production of powdered colourants.

2.4.6. Present day environment

The Erongo region comprises three different eco-zones: mountainous kopjes, plains, and river formations. These have to be taken into consideration as they influence the distribution of vegetal and faunal resources and subsequently the subsistence strategies of early communities (Wadley 1979). The current climate around Leopard Cave falls within the desert/Nama Karoo biome also known as the "Karoo shrubland" of northwestern Namibia, characterised by mountainous areas and open plains with mean temperatures of 18°C and summer rainfalls of around 100–500 mm with grass cover and mostly low shrubs that are drought-resistant (Mitchell 2002b, Sletten et al. 2013). The area becomes relatively hot in the summer with temperatures reaching 40°C. Thus, the vegetation around the site consists predominantly of sparsely distributed drought-resistant species (Table 1). Most are from the Capparaceae family consisting of very few *Boscia* and *Maerua* species, Fabaceae (Mimosoideae/Caesalpinoideae) characterised by a wide variety of thorny Vachellia/Senegalia species and *Adenolobus* and *Parkinsonia* species. Also present are *Combretum apiculatum*, *Terminalia prunioides*, *Commiphora* species, *Cyphostyemma* species on the mountain tops, an assortment of *Grewia* species, *Ximenia americana, Ziziphus mucronata* and *Vangueria infausta* and a few plants belonging to the Euphorbiaceae family. Poaceae (grasses), and other small herbs and creepers such as the wild melon, including *Citrullus lanatus* are seasonal. The Erongo escarpment is well known for its rock art sites even though they have not been explored after being brought to light in the 1950s by Abbé Breuil.

2.5. Toteng 1 (Botswana)

The Toteng sites are located around the village of the same name, between the extremities of the Okavango Delta and Lake Ngami not far from the confluence of the Kunyere and Nchabe Rivers (Robbins et al. 1998)(Figure 22). They

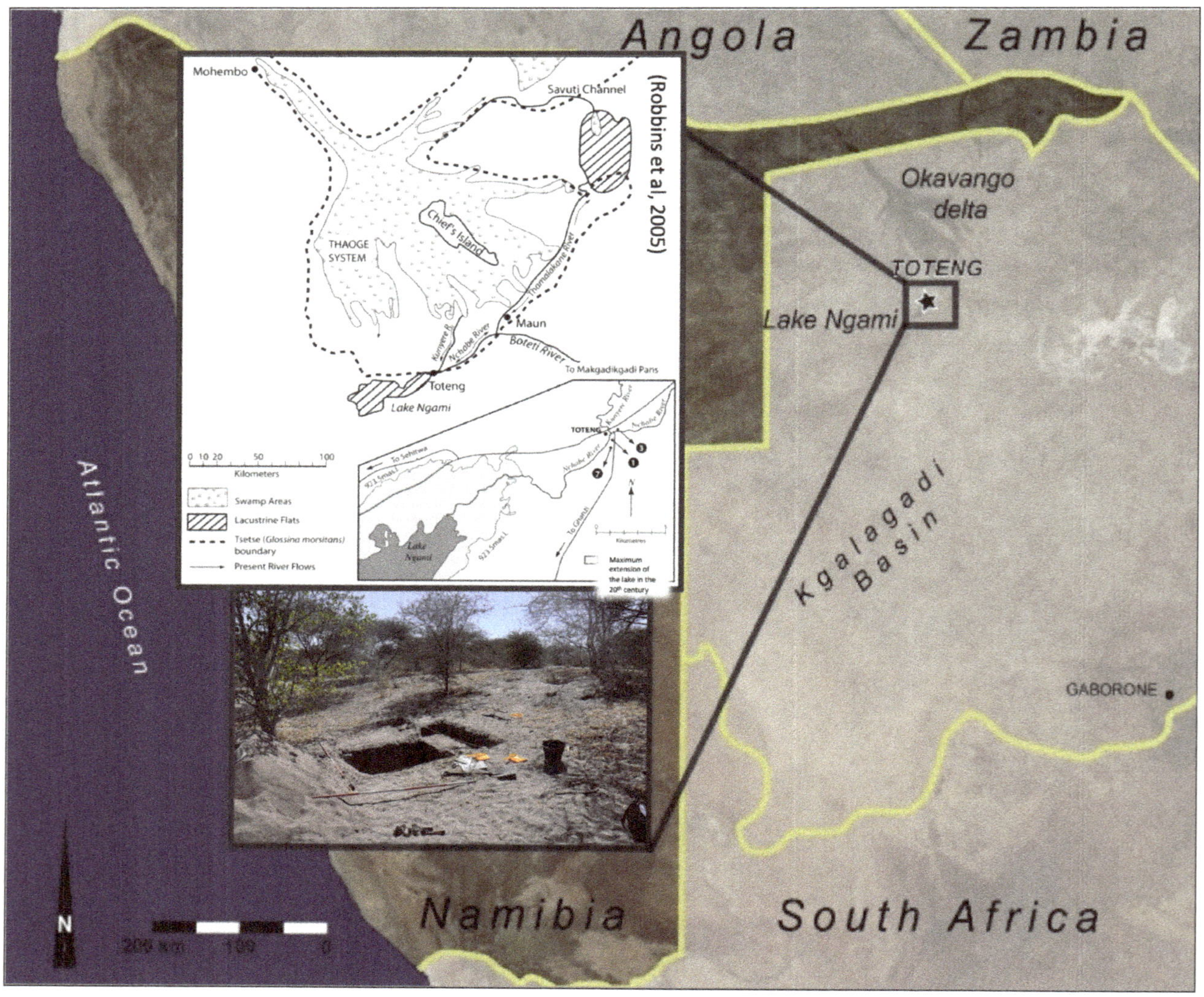

Figure 22. Location of the Toteng sites, in relation to Lake Ngami (☆) (after Robbins et al., 2005), with a view of the Toteng 1 sampling area in 2016.

were discovered in 1991 by an archaeological survey team led by A.C. Campbell (Campbell et al. 1992). Among the eight inventoried sites (labelled 1 to 8), two localities have been particularly studied: Toteng 1 and Toteng 3, with Toteng 1 containing the deepest archaeological deposits (Robbins et al. 2005, 2008). These two sites document human occupation layers attributed to the Later Stone Age, with hundreds of archaeological finds (microliths, bone arrows points, and ostrich eggshell beads). During these early investigations, refuse middens with Bambata pottery, together with bones from domestic animals (cattle and sheep/goat) were uncovered. An age of 1700 BP was established for the levels having delivered Bambata pottery and livestock remains on the basis of bulk radiocarbon ages. These levels are posterior to an earlier occupation without pottery or domesticated animals dated to 2640 BP (Campbell 1992).

These discoveries have revealed the archaeological potential of the Toteng sites for documenting the arrival of herding practices in the region. In 1996 an archaeological and environmental reconnaissance mission was carried out near Toteng to study the Holocene environments in the Lake Ngami Lake Sediment Zone (Robbins et al. 1998). In 2003–2004 a new excavation mission by an international team measured the chronology and stratigraphy of the occupations of Toteng 1 and 3, (Robbins et al. 2005 (Figure 23). This expedition improved the understanding of the sites' archaeological context. Additionally, in 2015 a brief expedition at Toteng 1 sampled in situ sediments for the study of the archaeobotanical remains that could frame the environmental conditions of the human occupation. This mission recovered the original stratigraphic section excavated in 2003–2004 and systematically sampled each layer (See chapter three). Thus, excavations at the site of Toteng 1 were of particular importance, because a stratigraphic sequence was developed spanning the period when early livestock first appeared in the area. Cattle and sheep appear circa 2020±40 BP (Robbins et al. 2008) making it a reference site for Botswana and, more generally, for the Southern African Later Stone Age period concerned with the spread of pastoralism and farming and the understanding of Late Holocene environments.

2.5.1. Stratigraphy and dating

Organic material from the site was first dated from earlier excavations where livestock bones were dated (Robbins et al. 2005) but more was added from the later excavations. The area was also sampled for the first ever optically stimulated luminescence (OSL) dating from the Toteng sites and the dates served to provide comparisons with AMS and bulk radiocarbon dates (Robbins et al. 2008). The location of the 2003–2004 excavation at Toteng 1 was chosen according to the results of the test pits realised in 1991. Among the tested areas, the squares labelled G, H, I, J, K and L were given priority as they had allowed the discovery of both the livestock bones and the deepest stratigraphic sequence. A total of eight squares were opened during the 2003–2004 excavation: F-L (Figure 23). A more than 1 m deep sequence (Figure 24), representing more than 6 m^3 of sediments was unearthed. It revealed four (approximately) horizontal stratigraphic units (SU 1–4), defined by their sediment characteristics (Robbins et al. 2008) as follows:

- SU 1: basal deposits of bedded pedogenic calcrete.
- SU 2: nodular silty-sand unit, bioturbated, with numerous animal burrows and roots.
- SU 3: black silty-sand.
- SU 4: organic-rich silty-sand.

The presence of the bioturbated sediments is clearly localised, particularly in SU2, where the incidence of darker sediment coming from SU3 is clearly visible. The chemical and textural analysis of Units 2–4 show a predominance of silt-sized particles with a clear presence of organic material (darker colours) and carbonate (lighter colours). Sand-sized sediments decreased from SU 2 to SU 4. A programme of dating helped to chronologically frame the archaeological sequence of Toteng 1. Both OSL and radiocarbon samples were analysed following the 1991 and 2003 expeditions (Campbell et al. 1992, Robbins et al. 2005, Robbins et al. 2008). They gave ages for the stratigraphic Units 2, 3 and 4 ranging from ca. 5000 to 600 cal BP. The OSL ages show that the sediments of Unit 2 accumulated from at least ca. 5000 to 4000 cal BP and those of Unit 2 were deposited from ca. 2000 to 600 cal BP (Table 7). The radiocarbon ages (Table 8) are consistent with these OSL dates. Particularly, three livestock bones (*Ovis aries* and *Bos taurus*) initially belonging to Unit 2 were dated and provided ages between ca. 2000 and 1350 cal BP (Robbins et al. 2005, Robbins et al. 2008).

In spite of the rich material evidence of Toteng and its significance to LSA in understanding the emergence of pastoralism through the analysis of dated stratigraphicaly-excavated deposits, what is absent from each of the excavations is any sampling for botanical remains. In 2016, therefore, the author carried out a stratigraphic sounding to add to the Toteng record (see chapter 3)

2.5.2. Archaeological remains

From the surface, the excavations revealed historic remains dating to the 19th century superposed on an LSA midden with both Bambata pottery and other artefacts such as beads and domestic livestock remains including positively identified cattle, sheep and goats. Even though occupation periods including surface layers through to Later Stone Age layers date from c. 3830 ±50 BP, the Bambata pottery associated with livestock has an average date of approximately 1710 BP at the Toteng sites. These contain a significant number of scatters of debitage, isolated bone points, decorated potsherds and well-preserved bones of large animals such as crocodile and hippopotamus as well as fish. Also present are lithic retouched microliths

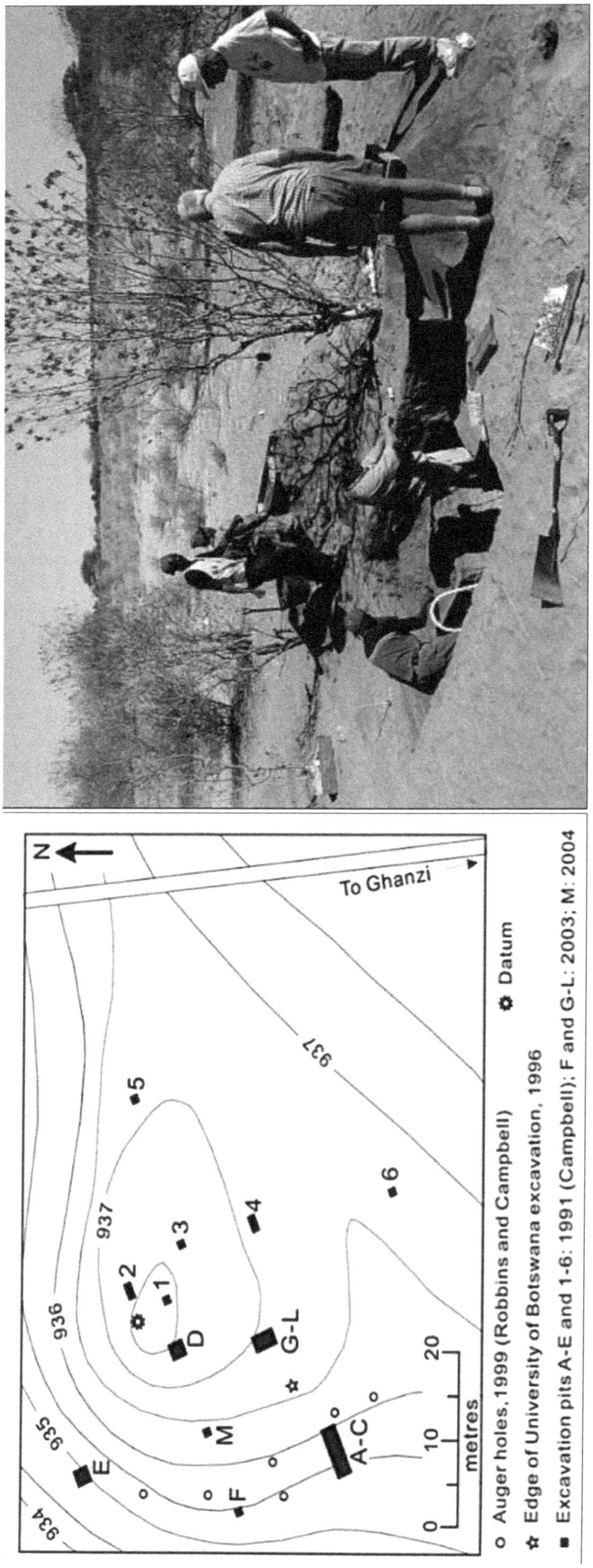

Figure 23. Location of the excavation areas (l) and view of the squares G-L excavation in 2003(r) (after Robbins et al., 2008).

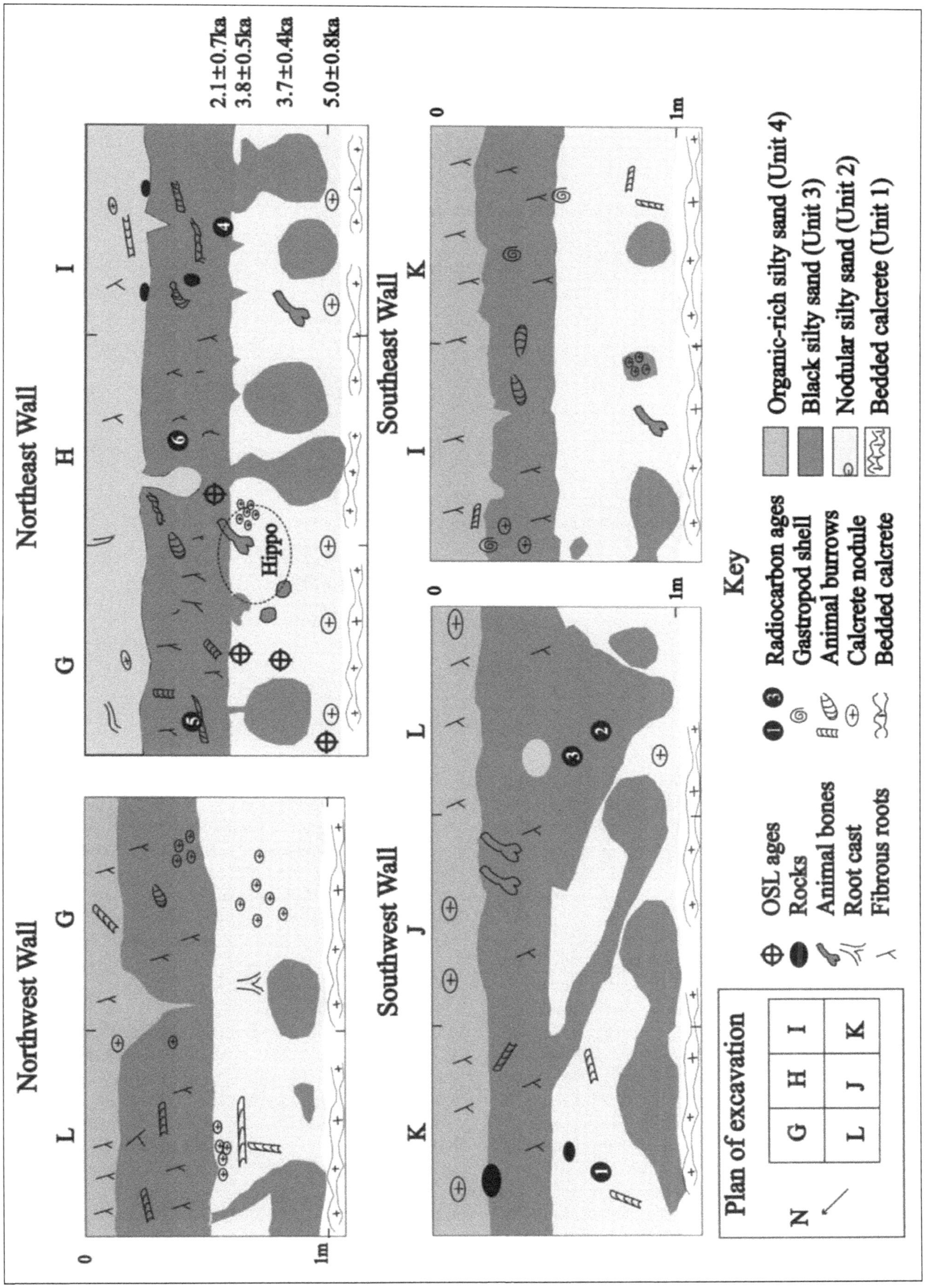

Figure 24. Stratigraphic sections from the six excavated squares and ages obtained for the Toteng 1 archaeological site (after Robbins et al., 2008).

Table 7. OSL ages from Toteng 1 sequence done in 2016

Lab #	Sample #	Unit & depth	U (ppm)	Th (ppm)	K (%)	Annual Dose (Gy/ka)	Paleodose (Gy)	Age (ka)
UGA03-117	TOT1 – 1	2 : 110 cm	1.1 ± 0.1	0.9 ± 0.4	0.21	0.7 ± 0.05	3.3 ± 0.05	5.0 ± 0.8
UGA03-119	TOT1 – 1a	2 : 78 cm	1.3 ± 0.1	0.7 ± 0.5	0.22	0.7 ± 0.06	2.6 ± 0.2	3.7 ± 0.4
UGA03-120	TOT1 – 1b	2 : 75 cm	1.4 ± 0.2	1.2 ± 0.6	0.23	0.8 ± 0.07	2.9 ± 0.3	3.8 ± 0.5
UGA03-121	TOT1 – 2	3 : 55 cm	1.2 ± 0.2	1.5 ± 0.6	0.22	0.7 ± 0.07	1.5 ± 0.5	2.1 ± 0.7

Table 8. AMS ages from Toteng 1. Calibrated using OxCal 4.3 and IntCal 13 curve (after Robbins et al. 2005, Robbins et al. 2008)

Lab #	Identification #	Square & depth	Age	Calibrated age BP (2 σ)
Beta 18669	O. aries, bone	Toteng 1, Sq. K, 75–80 cm	2020 +/–40 BP	2111–1835 cal BP
Beta 1904888	B. taurus, bone	Toteng 1, Sq. L, 70–75 cm	2070 +/–40 BP	2094–1890 cal BP
Beta 186670	B. taurus, tooth.	Toteng 1, Sq. L, 55–60 cm	1480+/–40 BP	1405–1283 cal BP
Beta 184609	Charcoal	Toteng 1, Sq. I, 55–60 cm	1560 +/–60 BP	1531–1307 cal BP
AA 58186	Charcoal	Toteng 1, Sq. G, 45–50 cm	1639 +/–33 BP	1565–1406 cal BP
AA 58188	Charcoal	Toteng 1, Sq. H, 40–45 cm	571 +/–31 BP	624–505 cal BP
AA 58187	Charcoal	Toteng 1, Sq. F, 35–40 cm	643 +/–31 BP	650–542 cal BP

Table 9. Toteng 1 artifact distribution within soil units from squares G-L. (after Robbins et al. 2008, Table 1)

Soil Unit	Lithics	OEB	Bone points	Undecorated sherds	Bambata sherd	Other decoration	Total
4 (0–35 cm)	221	6	5	52	3	4	291
3 (35–70 cm)	206	15	1	29	8	8	267
2(70–110 cm	198	6	7	9	2	0	222
1 (110–130)	37	1	1	0	0	0	39
Total	662	28	14	90	13	12	819

and a few grinding stone fragments and Bambata pottery usually recovered in association with domesticated animal remains. Other findings from the site include ostrich eggshells in the form of both finished and unfinished beads (Robbins et al. 2005, Robbins et al. 2008). Above (Table 9), is a summary of these artifacts as presented by Robbins et al. 2008.

2.5.3. Pottery

Apart from surface finds, Bambata pottery was found mainly within SU3 (35–60 cm) at Toteng 1. The first appearance of these decorated sherds in LSA deposits were dated to about 2000 BP in Unit 2. Bambata pottery is a typologically attributed type of decorated vessels characterised by very thin walls, fine decorations that include multiple bands, cross-hatching but are occasionally stamped and have incised lines on the neck (Huffman, 2007). They include jars and bowls, and they derive their name from the Bambata cave in Zimbabwe where they were initially discovered.

The Bambata pottery at Toteng was one of the initial few complete assemblages to be retrieved from a sedentary settlement while from neighbouring South Africa and Zimbabwe sherds are very few and fragmentary. They were then probably related to seasonal settlements and usually treated as "trait intrusions" (Huffman 1994, Robbins et al., 1998, 2008). Apart from Bambata pottery, traces of thick rim pottery with thick pointed bottoms associated with pastoral Khoi and short-necked painted vessels attributed to the Early Iron Age Bantu communities are also found.

2.5.4. Lithics

A total of 662 stone artefacts were recovered from squares G–L, some with traces of backing and retouch. The number of lithic tools varies between 0 and 4 per 5 cm level throughout the excavation. Among the retouched tools are segments/crescents, backed bladelets, double backed drills and a mixture of both a few large and small scrapers, all bearing Later Stone Age features.

Raw materials such as silcrete and different coloured types of chert, quartz/quartzite and chalcedony are found through the deposit of the G-L block. Chert and silcrete are found today in the local river around the Toteng area.

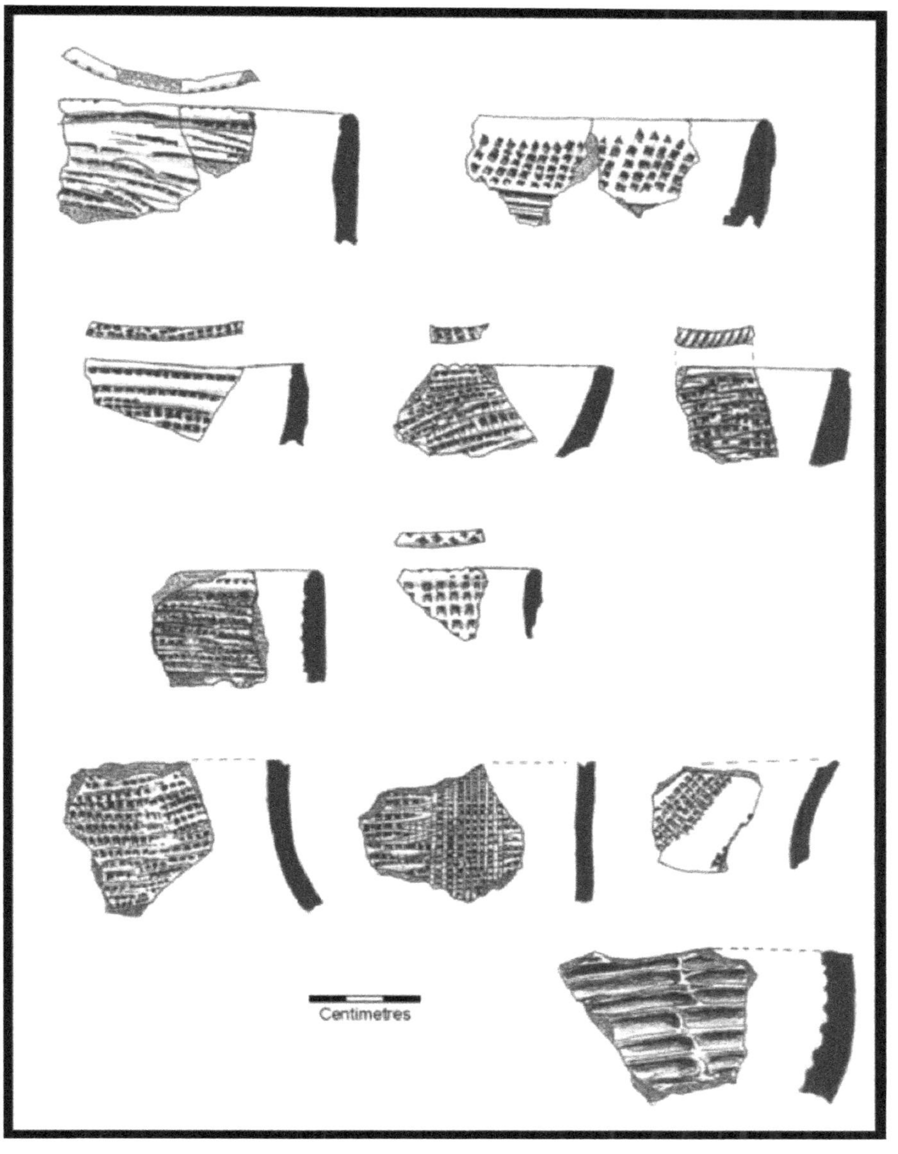

Figure 25. Bowl, possible bowl and possible jars of Bambata pottery from Toteng presenting comb-stamping, incisions and with red ochre (after Robbins et al. 2008).

2.5.5. Non lithic products

Bone remains include fragmented bone points and link shaft fragments throughout almost all of the squares excavated in 2003–4, confirming the 1991 findings (Morris et al. 2008). Ostrich eggshell beads were also found including unfinished beads and fragments detailing the different stages of bead manufacturing. These include rough-edged beads, raw undrilled bits, and partially drilled beads, complete and fragmented beads. The complete beads ranged from 6.5 to 2.5 mm in diameter, with a mean diameter of 4.3 mm with apertures with a mean diameter of 1.25 mm. Some bone point/link shaft fragments, typical LSA element, were also recovered from the excavations (both 1991 & 2003/4)

2.5.6. Faunal remains

A total of 929 bone and shell fragments from the 2003–4 excavations were identified from Toteng 1 revealing a wide spectrum of birds, bovids, rodents, suids, reptiles, caprines, some carnivores as well as fish. Bones of catfish, tortoise, monitor lizard, crocodile, hippopotamus, buffalo, lechwe, sitatunga, hartebeest, steenbok, duiker, sheep, and cow were thus recovered. All of the wild animals, as well as those recovered previously by Campbell, are known to have inhabited the Lake Ngami area historically.

Domestic animals were identified from both Toteng 1 and Toteng 3 with cattle (*Bos taurus*), sheep (*Ovis aries*), and possibly goat (*Capra*). Some of these remains have been directly dated, allowing the archaeologists to situate the appearance of domesticated animals in this region from around 2000 BP (Robbins et al. 2005, 2008). The cattle remains (N=3 for those surely Identified) from Toteng 1 provides the earliest known remains of this species in Southern Africa (Figure 27). They belong to SU 3 and 2.

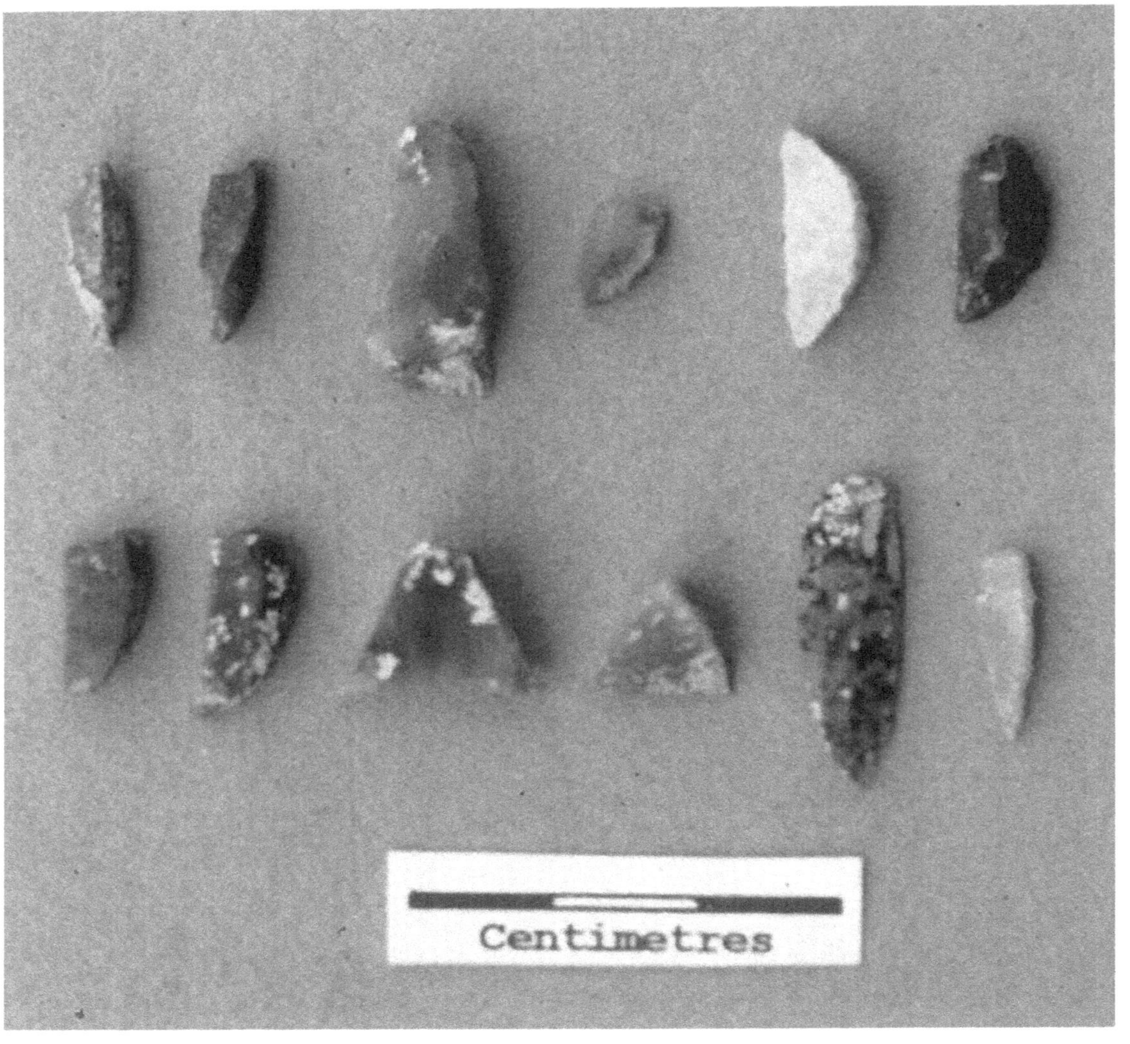

Figure 26. Lithics artifacts from Toteng 1, including backed drill and tools, as segments, trapezoid and distally backed points (after Robbins et al., 2008).

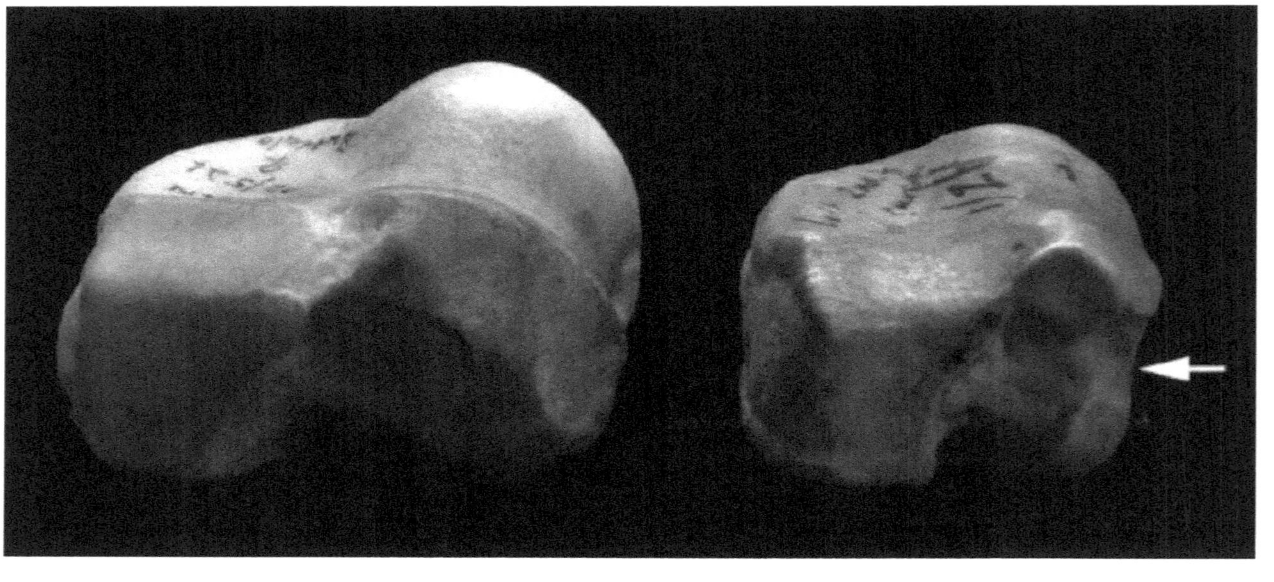

Figure 27. Comparison between buffalo and cattle carpals from Toteng (after Robbins et al. 2008).

Two positively identified sheep remains have been recovered in SU 2 and 4, whereas other possible caprines were also speculated for seven other bones. The sheep astragalus has been dated to 2020±40 BP and thus belong to the very earliest remains of caprines recovered in the whole of Southern Africa, also with those of Leopard Cave (Namibia, ca. 2200 BP), Spoegrivier Cave (South Africa, 2105 BP) or Blombos cave (South Africa, 1960 BP), which have all been discovered in Later Stone Age contexts.

Even though domesticates faunal remains are fascinating, finds such as a hippopotamus skull and a lion metatarsal also stand out. Most of the remains are from adult animals. However, there are a high number of fragmented bones preventing a precise identification. No slaughtering techniques could be deduced, and neither could the animals be sexed. The faunal assemblage is dominated by mammals still present in the vicinity of the site. The predominance of grazers shows that environmentally the conditions resemble the current ones.

Fish bones, mostly catfish, and tortoise-shell fragments were found in almost every 5-cm level of Toteng 1. The sitatunga (marshbuck) found at 95–100 cm and the lechwe (105–110 cm) are both wetland antelope, suggesting the proximity of swamps adjacent to the lake or rivers—a conclusion consistent with the previously mentioned OSL-dated sediments. A unique discovery at Toteng 1 was the intentional burial (perhaps for ritual purposes) of a hippopotamus head, the remains of which extended between 90 and 110 cm.

2.5.7. Lake Ngami as an important water source

Dated sediments from Toteng shed light on the nature of water fluctuations at Lake Ngami and finds like freshwater mammals and fish make the lake a very important source of information on how the lake influenced subsistence strategies and also to understand the palaeoenvironmental conditions and changes in Ngami during the Holocene. Imperative data from the sediments indicates that during these flow fluctuations, there was flowing or standing water near the Toteng 1 and 3 sites. Lake Ngami is found within the Ngami basin that is 55 km long containing a few outcrops and sand ridges. The lake itself is situated in the north part of the Kgalagadi desert in the vicinity of the Toteng village and Maun town in the northwest district of Botswana (Figure 22). The area has a semi-arid climate with annual precipitation of approximately 400 mm. The landscape around Lake Ngami area is characterised by aeolian and fluvio-lacustrine landforms associated with the Okavango Delta and its main water source is the Okavango River (Figure 22). Other water sources flow from the Thamalakane and the Kunyere rivers (Shaw 1985). Historical sources state that Lake Ngami has fluctuated from being dry to totally dry, recording significant seven dry phases since 1810. The latest drought the lake went through was recorded in

Figure 28. Present day Lake Ngami. Photo creditr: M. Mvimi.

1982 and it is reported that the surroundings were then littered by dying livestock. A completely dry phase was reported before 1951 and the lake was dry again in 1973–1974. Prior to this period from ca. 5500–4500 and again from 4000–3500 BP the lake was a permanent, extensive water body standing at an elevation of about 936 m. A noteworthy drop in the water level was recorded between 3700 and 2100 BP. It rose at around 2000 BP and later dropped again by 1200 BP. Even though the lake presents a series of changes, the Holocene in general is characterised by low water levels. During the Palaeo-Makgadikgadi stage, the Okavango Delta and the Ngami, Mababe and Makgadikgadi basins were linked together to form a single very large water body (McFarlane & Eckardt, 2006). The Makgadikgadi paleolake is now dry and Lake Ngami is its remnant (Cordova et al., 2017). From the archaeological record, apart from fish, large crocodiles were also found in the lake. Also present were such mammals as elephant, rhinoceros, hippopotamus, buffalo, giraffe, kudu, lechwe, reedbuck and impala. Birds such as wild ducks, pelicans, migrating flamingos and other birds are also present. The presence of these animals has made the lake an important area for hunting, fishing and foraging up to today (L. H. Robbins et al., 2008).

The lake basin contains ancient lake ridges and shorelines suggesting the presence of ancient lakes. The primary source of water for the basin and for Lake Ngami has always been overflow from the Okavango River delta. In the past 2000 years, people lived in the vicinity of the lake probably with seasonal occupations when the water levels in the lake receded. Historic accounts including those by David Livingstone mention the lake as an important source of fish with a wide variety present and fishing practices taking place. To date, the lake continues to be a source of food to the communities living around it (Robbins et al. 2005; personal observation 2016).

2.5.8. Present day landscape and vegetation

The evolution of landscapes in northwest Botswana has been studied far more effectively than other parts of the country and remain a source of geomorphological and palaeoenvironmental Holocene information for Southern Africa as a whole. The area around the Palaeo lakes, middle Kgalagadi and the Makgadikgadi consists of aeolian and lacustrine landforms of low relief covering an area associated with the Okavango Delta system (Thomas et al., 2003).

The vegetation type around the Okavango Delta is influenced by the ever-changing water levels resulting in complex environmental patterns. Occurring over time are mosaic-like vegetation patterns from permanent swamps, seasonally flooded types of swamps and grasslands to riverine woodlands and savannahs that are never under water (Ramberg et al., 2006) (Figure 29).

Figure 29. The current surroundings of Toteng, bottom left is part of Lake Ngami. Photo credit: M. Mvimi.

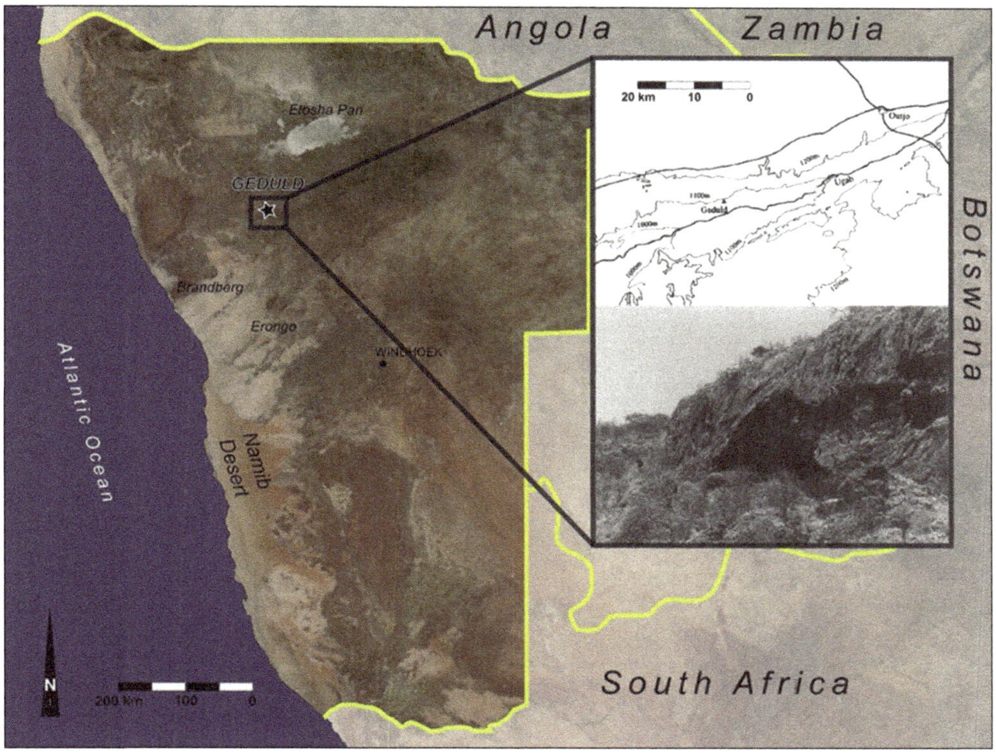

Figure 30. Location of Geduld (Namibia) (after Smith et al. 1995).

From observations around the site, the soils are aeolian in nature and the Nchabe and Kunyere rivers are dry during the winter months. Low lying ground cover comprised of grasses and other seasonal herbs and open woodlands dominated by a variety of Vachellia/Senegalias are common, as are familiar *Berchemia* and trees from the Combretaceae family (mainly *Terminalia*). The fauna found in the site, at least for the mammals and birds, are present in the area today, mostly observable around Maun town as well as in the wet Okavango Delta.

Today Pastoralism is practised in the Toteng village and people keep sheep, cattle, goats, cats and dogs among other small stock. Horses are present and used as a mode of transport, and include stray horses (known as Kgalagadi horses) introduced during the colonial period (Cordova et al., 2017).

2.6. Geduld (Kunene, Namibia)

Geduld is situated at (20.17S 15.50E) in the Kunene region of northwestern Namibia, on the northern bank of the Ugab river, 35 km southwest of the modern town of Outjo (Figure 30). The vegetation around the site falls into the Mophane, thornbush, and Damara thornveld and mountain savannah biome. The area is a summer rainfall area receiving ca. 400–1000 mm of rain annually.

2.6.1. Stratigraphy and datings

Excavation at the site of Geduld took place between 1978 and 1986 (A. B. Smith et al., 1995). A first test pit was excavated measuring 1x1 m, Square B5 and later eight more squares were opened to extend these squares namely B4-B9 and C5-C7 which are a horizontal extension to squares B5 and B7, giving a 6 × 1 m with a 2 × 2 m at B6-B7 and C5-C7.

More than thirteen stratigraphic layers have been identified, mainly composed of soft ash intersected by some (burnt) organic bedding material and dung layers. Some local bioturbation was also visible. Consequently, the integrity of the infilling has been questioned by J. Kinahan (1995) who highlighted the presence of possible disturbance which would possibly disconnect livestock remains that easily could have migrated down the sequence from their original position.

Radiocarbon dates on organic material show that the site occupation ranges from 2300±50 BP (L13, z=72cm) to 800±50 BP (L1A, z=4cm) (Smith et al 1995 ;Smith 2008). New dates (2018) from charcoal from square B9, Layer 3, gave the age of 1676±25 BP.

2.6.2. Material cultures

The excavation produced a variety of material culture including a ceramic sequence dating to around 2000 BP, ornamentations made of ostrich eggshell, a stone tool assemblage organic material like charcoal and seeds, and ash deposits. During excavation, a large quantity of lithic material, amounting to more than 23 000 pieces, was recovered (Smith et al 1995, Smith 2008). The lithic artefacts consist of much debitage material, including

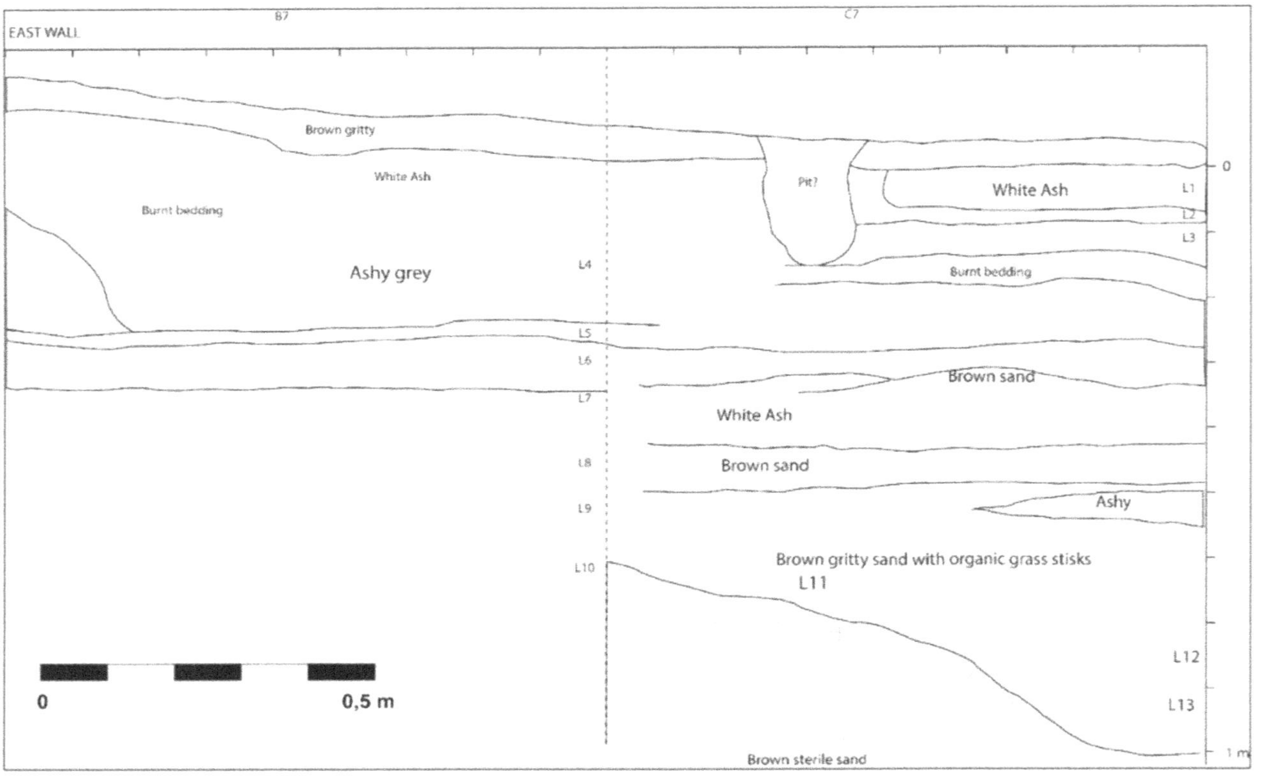

Figure 31. Stratigraphic sequence of the Geduld excavation (after Smith et al., 1995).

Table 10. Radiocarbon ages obtained at Geduld (after Smith et al. 1995)

Lab#	Year	Square	Layer	Depth (cm)	Nature	Conventional Radiocarbon Age BP
Pta-4416	Smith et al., 1995	C7	L1A	4	Charcoal	800 ±50
Pta-4416	Smith et al., 1995	B6	L5	26	Dung	1790 ±80
Pta-4419	Smith et al., 1995	B5		30-40	Charcoal	1790 ±50
Pta-2720	Smith et al., 1995	C6	L7	36	Charcoal	1980 ±50
Pta-4413	Smith et al., 1995	C7	L8	46	Charcoal	1970 ±40
Pta-5875	Smith et al., 1995	C7	L9	51	Charcoal	2090 ±45
Pta-5871	Smith et al., 1995	C7/B7	L10	55	Charcoal	2040 ±45
Pta-5873	Smith et al., 1995	C7	L11	63	Charcoal	2110 ±60
Pta-4414	Smith et al., 1995	C7	L13	72	Charcoal	2300 ±50
UBA-38618	This study, 2018	B9	L3	–	Charcoal	1676±25

cores and retouched stone tools. Microliths including segments and backed blades, scrapers and an adze, and a concave stone that could have been a grinding stone or an anvil were also discovered. The microlithic remains conform to the Late Stone Age period as they are similar to those found in other sites of the same period in Namibia. In addition, customised bone items, both finished and unfinished, burnt bone, notched bone, burnt ostrich eggshells and tortoise shell remains were found in squares B5, B6 and C7. Ostrich egg shells beads were analysed by Jacobson (2010) comparing the diameter of the OEB and concluding that there is a notable increase in the shell diameter from the earlier to the later periods. Together with the OES ornaments, both mollusc shell and seed beads were found. The seed beads were found connected by a fibre string. Decorated ceramics dominated by thin-walled vessels were also found at the site. In square B7, level 2, and B7, level 4–7 a small triangular blade and another identified iron object were discovered.

2.6.3. Faunal remains

The excavation recovered a mammalian assemblage consisting mainly of bovids (78 percent, including Damara dik-dik, hartebeest or steenbok), reptiles and birds as well as marine animals (seal) (Table 11). Micro-mammalian fauna, reptiles and amphibians were also identified from almost all levels. Particularly, among the small bovid assemblage, some domestic stock remains dating to 1800 BP were identified. Sheep bones were recovered from Layer 4 (astragal, phalanx, teeth) from different squares. Charcoal from this same layer has been radiocarbon dated to 1790± 50 BP, confirming the early appearance of domesticated animals in the sequence. Most of these remains have not been directly dated, which may limit including Geduld in the first southern African herding practices debates.

Table 11. Faunal spectra from Geduld (after Smith et al., 1995)

	Surf/L1	L2	L3	L4	L1-4	L5	L6	L7	L8	L9	L10	L11	L14	TOTAL	%
MAMMALS															
seal								1			1			2	0,9
hare			1					1		1				3	1,4
cf. hare		7						2	1					10	4,7
cf. genet	1													1	0,5
cf. mongoose	2			1				1						4	1,9
equid				1			8	3						12	5,6
dassie	1		3	2	1								1	8	3,7
cf. dassie			4		1									5	2,3
rhino	1			1										2	0,9
hartebeest														1	0,9
dikdik	1													1	0,5
grey duiker						1		2						3	1,4
sheep	1			5										6	2,8
Bovidae sp. Indet:															
small	2	4	8	10		5	7	22	5					63	29,3
medium	1	2	14	9		6	6	8	1					47	21,9
medium large	4	3	7	5		9	3	2		3			2	38	17,7
large								1	1					9	2,8
TOTAL														**215**	**100**
BIRDS															
francolin	1													1	
hornbill	?1													1	
common kestrel			?1											1	
Cape turtle dove			1											1	
helmeted guinea fowl						2	2						4	4	
gen. et sp. Inidentified				1										2	
TOTAL:														**10**	
REPTILES															
bullfrog	1													1	
tortoise	29	44	20	48	9	38	36	120	17	3	7	4	2	377	
snake	1		3											4	
reptile sp. indet.	5	5	2	2			3						1	18	
unident. bone frags (mass in grams)	795	495	540	1099	108	975	733	1363	172	55	21	13	22	6391	

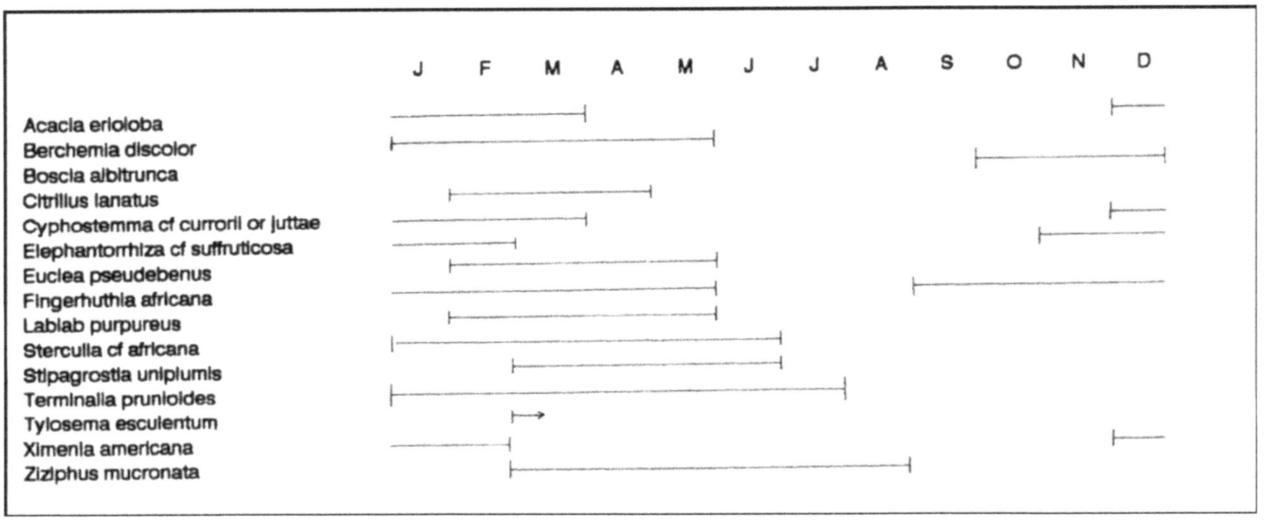

Figure 32. Fruiting seasons of plants identified from Geduld (after Smith et al., 1995).

2.6.4. Botanical remains

Most botanical identifications were those of seeds. Jacobson et al (1995) found it difficult to identify the rest of the plant remains. From observation, these unidentifiable plants remains include twigs, bark and sometimes roots from trees.

Judging from the flowering patterns of the plants, most of the fruit is produced in late summer. They should be regarded as food plants that were also used as medicine or infusions though mostly as firewood.

Geduld, or archaeological assemblage acquired therefrom, is too small to draw conclusions on cultural transitions that may have occurred in the region. Whether the inhabitants of Geduld formed an entirely pastoral society or they remained a hunting community in contact with herders from whom they occasionally acquired domestic animals is unclear. The presence of plants such as *Lablab purpurus* and *Ricinus communis* which are commonly used for livestock care and feed suggest that the inhabitants of Geduld practised animal husbandry, and indeed owned livestock. The presence of iron tools raises questions about the acquisition of metal items and the identity of the people as in the region, in that iron smelting was introduced into southern Africa by Bantu speakers sometime after 2000 BP. The same ambiguity applies to the presence of pottery: was it produced locally or acquired through exchanges of products with other groups? Although Geduld has yielded a limited quantity of material from a single excavation, it serves is a good reference in addressing questions on the emergence of pottery and livestock. Pottery in the area as well as the earliest indications of food production precedes livestock. As is the case with other LSA livestock remains in the region (Toteng, Spoerivier etc.) they are very limited, making it difficult to determine the connection of livestock acquisition and development. As far as unchanging nature of lithic technologies during periods of pottery and later livestock introduction Geduld is useful alongside other sites for understanding the identity of LSA communities.

3

Archaeobotanical Material and Methods

3.1. Description of the study material

The material from the three sites concerned with this study – Toteng, Leopard Cave, and Geduld – consist of macrobotanical remains that are plant elements normally distinguishable by the naked eye (usually > 1 mm) as opposed to microbotanical remains such as pollen and phytoliths that require a microscope to be observed, sorted or identified. Most of the remains correspond to wood fragments while other elements from the vegetative part of plants (twigs, fragments of bark and herbaceous stems) are also present in the samples along with seed and fruit remains. Table 12 lists the different categories of plant remains that have been analysed from the three sites.

While the majority of ligneous remains were found in a carbonised state, most likely due to the use of wood as fuel, seeds and fruits were more often preserved by desiccation. Indeed, the arid conditions prevailing at all three sites situated at the edge of the Kgalagadi desert allow the conservation of organic material in a dry state, a type of preservation that is otherwise rare on archaeological sites.

In Leopard Cave, a small number of seeds belonging to two taxa (*Grewia* sp. and *Croton* sp.) have been preserved by mineralisation. This type of conservation implies that the original organic material has progressively been replaced by mineral components. Two different processes can account for this (Messager et al. 2010). In some plants, the replacement by minerals of specific tissues, for example, fruit stones (botanically endocarps) can occur naturally while the plant is alive, a phenomenon called bio-mineralisation. Bio-mineralisation has been shown to occur in specific taxa such as hackberry (*Celtis*) or in nutlets from the *Boraginaceae* family. In other cases, it occurs in an environment rich in mineral salts responsible for the mineralisation of organic material. In archaeological contexts, this is common for example in latrines or wells where humidity and different mineral components such as phosphates and calcium occur together. In both cases, the inorganic nature of the replacement minerals means that the structures become resilient to decomposition, which is recognised as the reason for the abundance of certain taxa at many archaeological sites (Shillito & Almond, 2010). Given the conditions at Leopard Cave, the first type of preservation –bio-mineralisation – seems more likely but due to a lack of adequate literature on this phenomenon within *Grewia* and *Croton* species, a definite conclusion cannot be drawn.

The preservation of botanical macroremains is comparatively good on all of the studied sites. Charcoal and wood pieces occur in different sizes sometimes measuring up to 3 cm as it is the case in certain hearths in Leopard Cave. Most of the seeds were found intact even though some were split into halves or more fragmented (Figure 33).

However, differences between the three sites have been observed. The two rock shelters (Leopard Cave and Geduld) are situated in the Erongo escarpments and, except for bioturbation mostly by burrowing rodents; they are protected from other erosional factors that will affect open-air sites such as Toteng. On the latter site, the environment is characterised by aeolian and fluvio-lacustrine conditions and the archaeological settlement is exposed to flooding, wind erosion as well as animal trampling that might hamper the preservation of fragile organic material. Subsequently, charcoal pieces are less numerous and more fragmented at Toteng than at the other two sites, where botanical remains are in general both more abundant and better preserved.

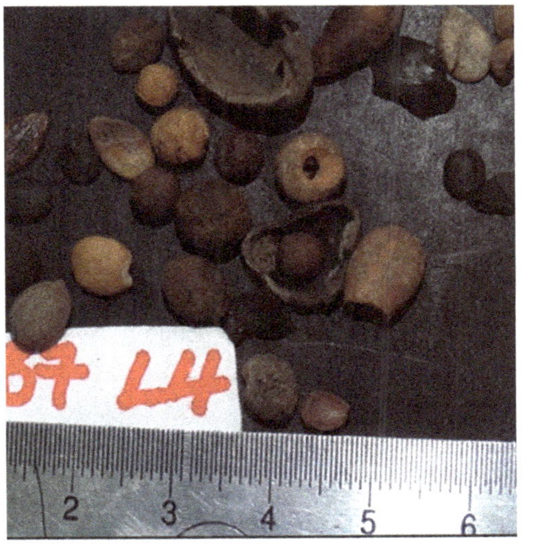

Figure 33. A mixture of carbonised and desiccated fruit remains from Geduld (square B7, level 4). Photo credit: M. Mvimi.

Table 12. Categories of plant remains studied from each of the three sites

Site	Charcoal	Dry wood	Charred seeds/fruits	Dry seeds/fruits	Mineralised seeds/fruits
Leopard Cave	X		X	X	X
Geduld	X	X	X	X	
Toteng	X				

Still, this type of rock-shelter sites is very rare in southern Africa given the generally flat topography of the region.The fact that the desiccated botanical remains may look very similar to modern plant parts raises the question about the authenticity of their suggested age. The literature however reveals that it is common to find exceptional preservation conditions at earlier or contemporary sites located in semi-arid conditions in Southern Africa (Prior & Williams, 1985; Sievers, 2006). The same applies to the Sahara (Dunne et al., 2016) and probably in desert and semi-desert conditions where the biodegradation processes have severe limits. When seeds from Geduld and Leopard Cave were C14 dated, results confirmed that appearances can be deceiving. These included seeds from layers where previous C14 dates were obtained on charcoal in determining the age of the site. These dates did indeed attribute to some of the "modern-looking" seeds (square P7, level 150–160 cm) an expectedly old date: 1991±33 BP. In another case, a seed found in a lower level in Leopard Cave (square O8, level 200–205 cm) the C14 dating indicated a recent date (88±30 BP), suggesting an intrusion by burrowing rodents regularly observed during excavation. This material was excluded from the study assemblage.

3.2. Sampling methods

The methods for sampling of macrobotanical remains applied at the three sites differed somewhat. At Leopard Cave, systematic sampling for both animal and plant remains was carried out from the beginning of the excavations in 2009, under the direction of D. Pleurdeau (MNHN). At Geduld, all the botanical remains that were examined had been collected by the American team led by Leon Jacobson during excavations between 1978 and 1986 and stored at the National Museum of Namibia, Windhoek. At Toteng 1, no archaeobotanical remains had been retrieved during the excavations, therefore a second visit to the site in 2016 was organised in order to conduct systematic sampling from the different layers.

3.3. Leopard Cave

Botanical sampling was made of remains collected since the excavations started in 2012. Between 2013 and 2018 the author participated in the yearly fieldwork in order to collect supplementary wood and seed remains as well as to create a local reference collection (see below). This study is concerned only with the material collected in 2012 and 2015, with the remaining samples to be examined as part of future studies. During the fieldwork in 2015, botanical remains were collected from squares P7 and O8 at the intersections of sub-squares a/b-c/d (Figure 34, Figure 35). Soil samples were collected from the stratigraphic sequence revealed in square O8 (Figure 35). These were chosen from the different layers that constitute a more than 1 m high W-E profile and correspond to a period spanning 2190±40 to 3250±40 BP according to the C14 dates obtained from charcoal fragments. The samples came from dark to greyish silty and/or ashy deposits primarily containing vegetal remains as well as occasionally granitic fragments distributed along the profile, from maximum depths of 220 cm to 140 cm. Sediments were collected from more or less circular zones (diameter 12–26 cm) dug into the sediment profile with the help of a trowel and then contained in 2 l labelled plastic bags. A total weight of 23 kg of soil was collected in this way.

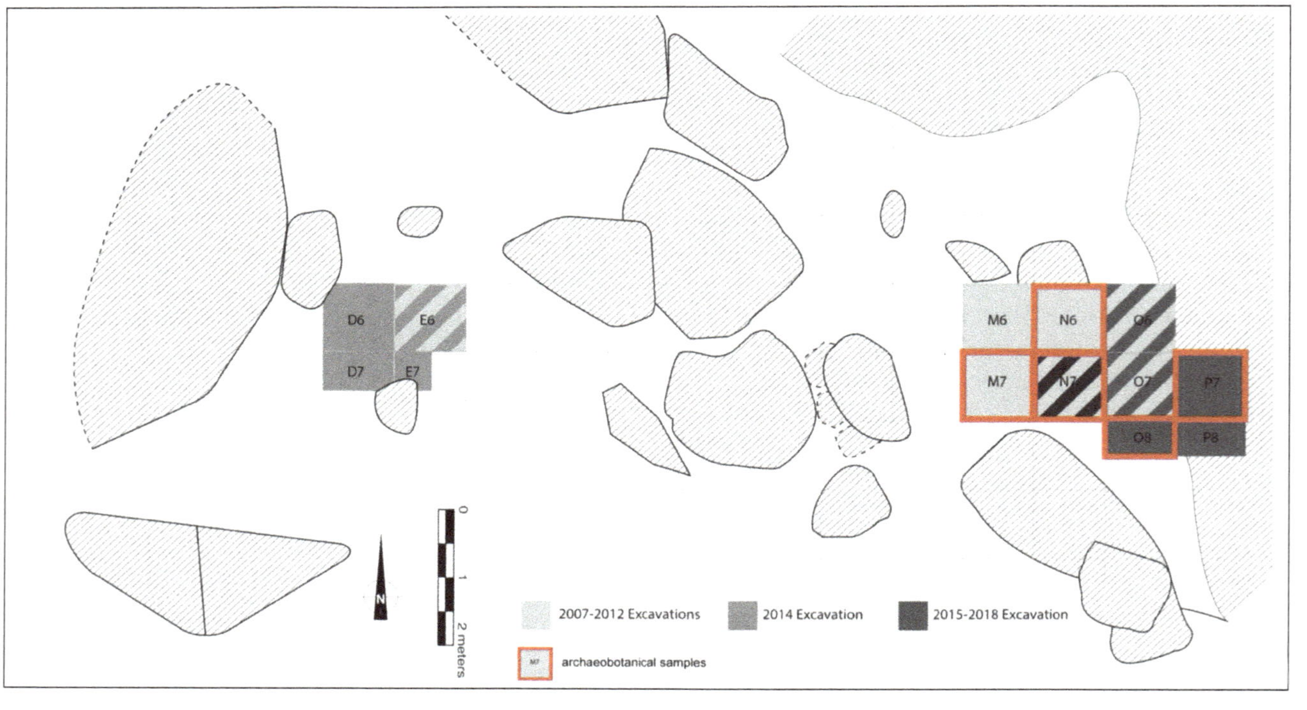

Figure 34. General overview of the excavation area at Leopard Cave, red squares indicate zones that were sampled for archaeobotanical remains. Authors: D. Pleurdeau/M.Mvimi.

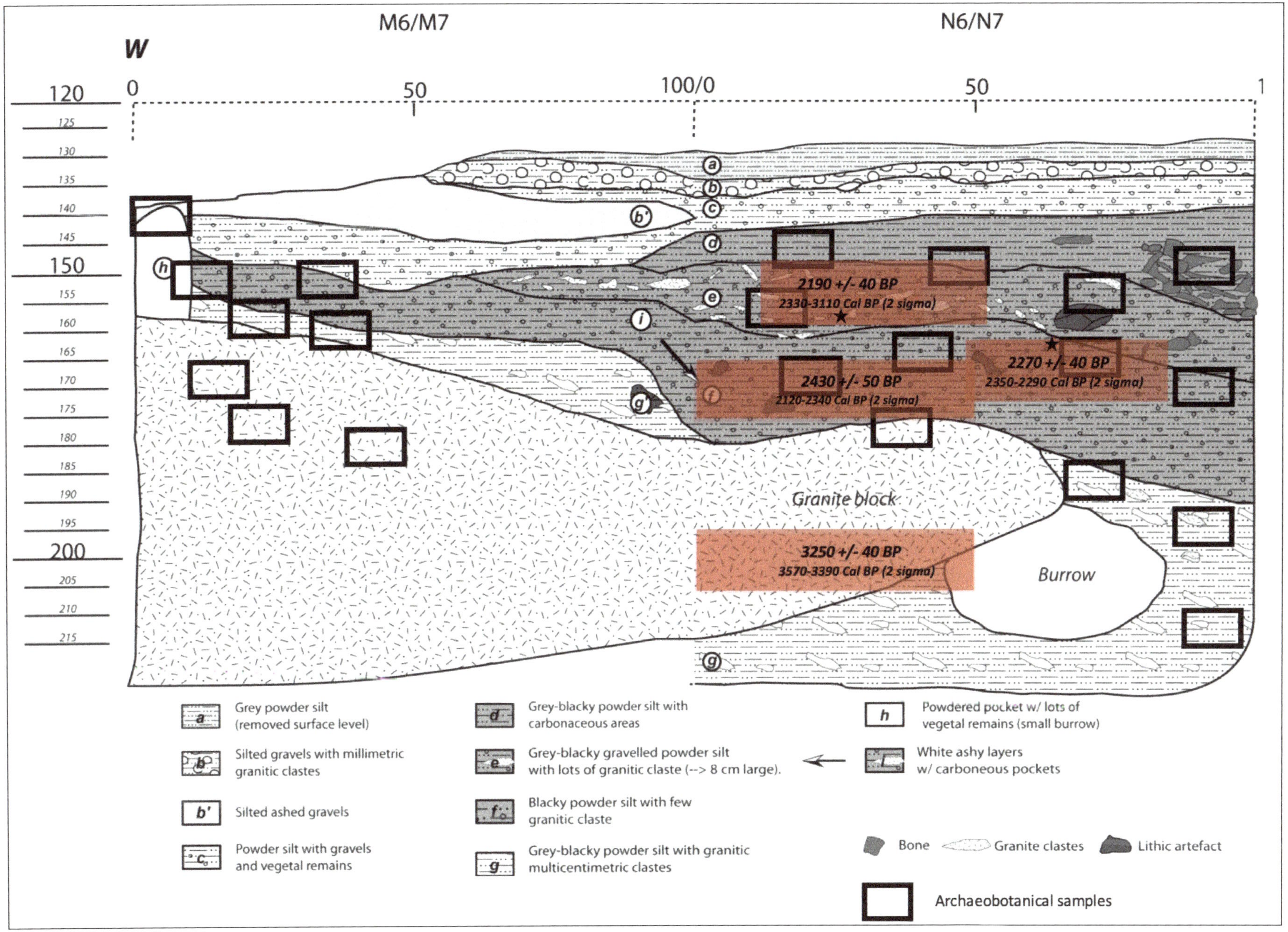

Figure 35. W-E profile of square O8 in Leopard Cave. Author: D. Pleurdeau.

In parallel, plant remains were collected during the excavation of square P7 either picked by hand or sampled in the form of sediment. Thus, large charcoal pieces were extracted manually from two concentrations of large-size charcoal pieces found in situ and interpreted as hearths. The first of these is named Hearth 1 (*Foyer* 1 or F1) and corresponds to a circular structure situated at a depth of 180 cm. A similar concentration but with fewer charcoal fragments was found adjacent to F1 at a distance of 23 cm and was named F2. Additional charcoal samples were collected within sediments from ashy deposits. Both these soil samples and those collected from the stratigraphy (see above) were dry-sieved with a sieve mesh of 10 mm. A large fraction was sorted by the naked eye. Granite fragments were eliminated; lithic pieces, charcoal, seeds/fruits and faunal remains were sorted, contained in bags and labelled. The fine fraction was collected, and botanical remains were extracted by bucket flotation carried out on site. Very few seeds and charcoal remains were found in the fine fraction of samples.

3.4. Geduld

Botanical remains from Geduld were collected during the excavations conducted at the site from 1978 to 1986 under the direction of L. Jacobson. They originate from different squares and levels named during the archaeological fieldwork (Table 13) and according to the radiocarbon dates obtained span a period from 2300±50 BP to 1676±25 BP. A total of 33 samples of botanical remains were exported to Paris for study thanks to the kind permission of Ms Fousy Kambombo, the curator at the time.

Resulting publications (Yates et al. 1995) provided no precise information of how the plant remains were collected. The samples stored in the National Museum were packaged in small, labelled plastic and paper bags that incorporated site information and the provenance of the materials. The smaller bags were stored within larger boxes corresponding to the different stratigraphic units. When not separated into labelled bags, sometimes they were found mixed with other archaeological material, including coprolites and other plant remains. The majority of the charcoal pieces and seed remains were complemented by soil sediment which suggests that dry sieving may have been practised.

Most of the seed and fruit remains, found both in a carbonised and desiccated state, had already been the subject of a preliminary study (Yates et al. 1995) allowing the identification of 24 botanical taxa. However, they were included in this study in order to verify identifications and to compare the results with those of the charcoal study.

3.5. Toteng 1

In the archaeological Toteng collections kept at the Botswana National Museum in Gaborone, archaeobotanical material is almost non-existent, because previous studies apparently did not include this type of study. Given the importance of the site to the study period and especially the presence of the earliest domesticated animals attested in Southern Africa so far (see chapter II) it was vital to visit the site in 2016 and open a stratigraphic trench that would allow the sampling of plant remains.

The original excavation was located after an hour of navigation with a GPS and thanks to a member of the National Museum of Botswana who was present at the time of the 2002 excavations whose precise location the team managed to locate. Once the previous excavation trench had been delimited, an excavation area of 2 × 1 m was divided into two $1m^2$ squares named N and O. The sampling was done on the west-facing section on the half intersections (50 cm each) of each square (Figure 36). Nine samples were obtained, the highest at a depth of 20 cm and the lowest at 110 cm. Spits with a horizontal diameter of

Table 13. List of samples studied from Geduld

Square	Level	C14 date layer	Description
B5-6	L1/4	/	Bedding
B6	L5	1790±80 BP	Ash/bedding
B6	L7	/	Ash/bedding
B6	L6	/	Ash/bedding
B7	L5	/	Ash
B7	L7	/	Ash/bedding
B7	L6/L7	/	Ash/bedding
B8	L3	/	Gritty/ash
B8	L4	/	Dung/ash/hearth
B8	L1/4	/	Dung/ash/hearth
B9	L1/3	1791±50 BP, 1676±25 BP	Dung/ash/burrow
C6	L8	/	Bedding/ash
C7	L8/L10	/	Ash
C7	L11/14	2300±50 BP, 2110±60 BP	Sand/grass

Figure 36. Excavation of Toteng 1 in 2016. Photo credit: D. Pleurdeau.

10 cm and a depth of 20 cm were selected for sampling. A total of 18 kg of sediment was collected in this way and collected in 5 l mini-grip bags before transportation to the National Museum where the samples were treated by flotation. During excavation and sieving, some charcoal pieces and seeds were manually retrieved. At the National Museum, the sediment was first dry-sieved through a 2 mm mesh and larger charcoal and seed fragments were picked by hand. The sieved sediment was kept and later bucket flotation was done. The archaeological sediment was poured into 10 l buckets filled with water. The floating charred remains were recovered by sieving through a fine-meshed textile that was then hung to dry in the sun. Finally, the dry botanical material was wrapped in aluminium foil, labelled and packed in small mini-grip bags. In total 198 charcoal fragments were recovered in this manner. The site was very poor in seed remains producing only 9 *Vachellia/Senegalia* seeds. Moreover, all were retrieved from the uppermost levels of the excavation (0–20 cm) and were considered to correspond to modern pollution. The presence of an *Vachelia tortilis* tree very close to our trench probably explains the presence of seeds in the uppermost layers.

3.6. Laboratory methods and botanical identifications

Identification of botanical remains was carried out in two laboratories specialised in archaeobotany: the laboratory hosted by the Institut Català de Paleologica Humana i evolucio social (IPHES) in Tarragona, Spain and the laboratory of the research unit 7209 (*Archaeozoology,*

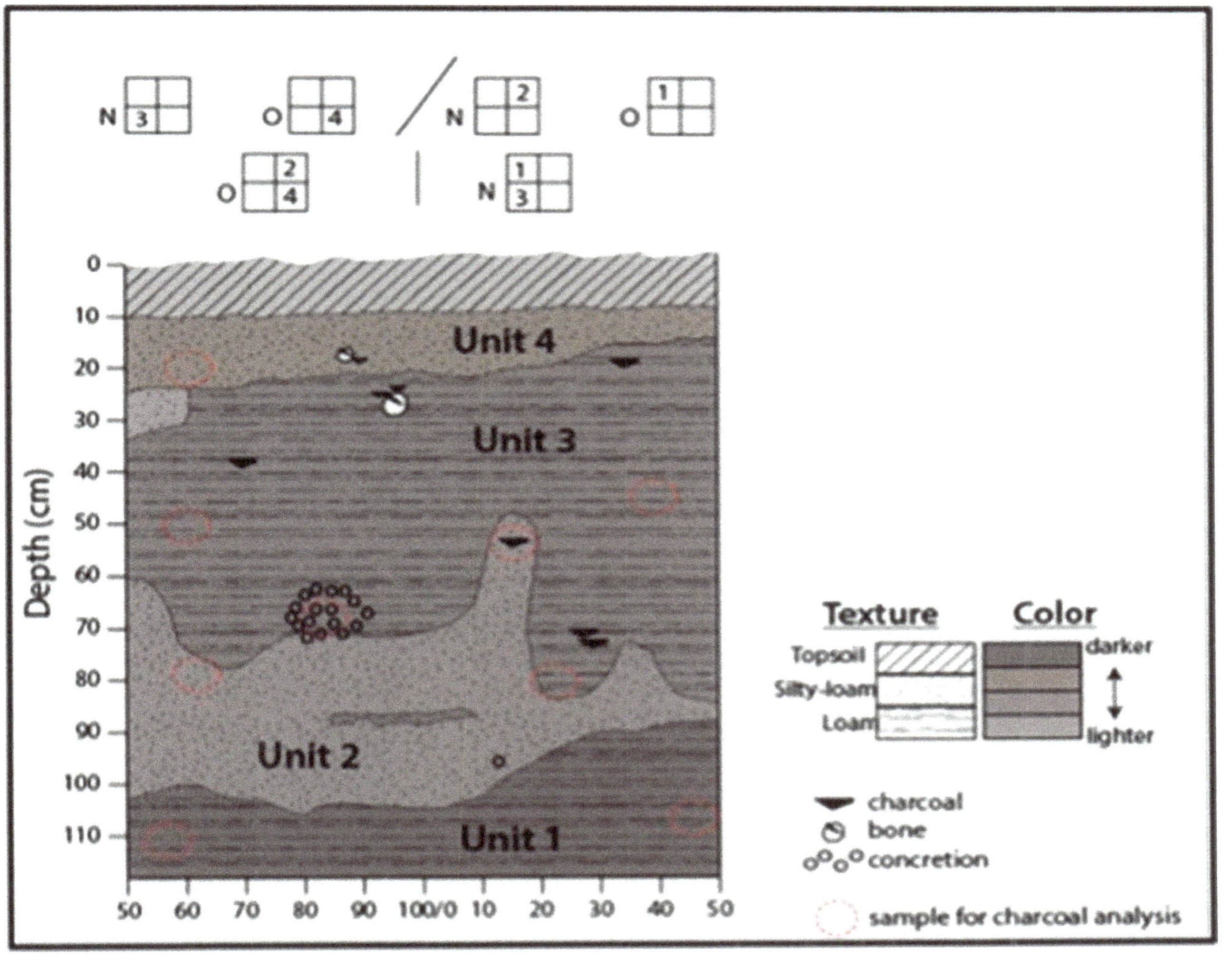

Figure 37. Drawing of the stratigraphy revealed in 2016 (r) and section from the 2002–03 excavation (l), (after Robbins et al. 2008).

Figure 38. Use of a reflected-light microscope for charcoal analysis in the archaeobotanical laboratory of the National Museum for Natural History, Paris. Photo credit: M-P Ruas.

Archaeobotany: Societies, practices and environments) in the National Museum for Natural History, Paris (Figure 38). The laboratory work was carried out alternatively under the supervision of Dr E. Allué in Tarragona and Dr M. Tengberg in Paris. Moreover, two visits to the Institut für Archäologische Wissenschaften at the Goethe-University in Frankfurt am Main allowed this study to benefit from the special expertise in African archaeobotany that has been developed by the team directed by Prof K. Neumann.

3.7. Charcoal analysis

To prepare the charcoal for identification the charcoal pieces were fractured with the help of a razor blade according to the three sections of the wood: transversal, longitudinal radial and longitudinal tangential. The next step was to observe the anatomical features of the archaeological wood under an Olympus BX51TRF incident light microscope at different magnifications (x50-x1000). Salt in a Petri dish placed on the microscope stage was used to hold the charcoal in place during analysis. Chosen charcoal pieces were prepared for photography and pictures were realised using a JEOL (x10-x20000) JSM-5000 Neoscope Scanning Electron Microscope.

In order to identify the wood samples, the protocol followed is that of the International Association of Wood Anatomists (IAWA) (IAWA Committee, 1989). For comparison, wood anatomical atlases and anthracological catalogues (Neumann et al., 2001) as well the Inside Wood Internet database (InsideWood, 2004 onwards) were utilised. Although these references have been of great help, none is specialised on taxa from the study region.

Reference collections held in the archaeobotanical laboratories in Paris and Tarragona were consulted but did not always contain the representative species. The collection in the Natural History Museum in Paris is for example oriented towards the arid regions of the Middle East. Even though some of the genera are common to the Southern African region, (*Vachellia/Senegalia, Maerua, Boscia…*) precise species are not. In this respect, several visits to the specialised laboratory at the Goethe University in Frankfurt, where there is a large reference collection of African woods, as well as the advice from the specialists working there, were particularly advantageous. Still, even at Frankfurt, reference collections and research projects are predominantly oriented towards West Africa, albeit some investigations haveconcerned the study region concerned with here, as explained in chapter 1.

While the diversity of taxa at the three sites is comparatively low due to the arid environment (see results and discussion) their identification was in many cases problematic. *Vachellia/Senegalia*, the predominant genus at all three sites, comprises a large group of species (in Southern Africa) that show very similar anatomical characteristics all the while presenting a large intra-specific diversity (Neumann et al. 2001). Apart from a few species, for example, *V. nilotica* that show particular characteristics (in this case, clearly defined circumvascular parenchyma), it is not possible to push the identification further than to the genus level (*Vachellia/Senegalia* sp. or when several types are observed: *Vachellia/Senegalia* spp). Moreover, other genera, in particular, *Dichrostachys,* equally of the Fabaceae family, could sometimes not be distinguished from *Vachellia/Senegalia*.In these cases, the two were grouped together as *Vachellia/Senegalia/ Dichrostachys.* To tell apart the anatomically close genera *Boscia* and *Maerua* from the Capparaceae family was another challenge and often these two had to be grouped together as *Boscia/Maerua*.

To address these difficulties, the 'type' concept (Ekblom et al. 2014) was applied. This model uses a type to represent a morphological or, in the case of wood charcoal, an anatomical category that is distinguished from others by a single character or a significant character combination. Therefore a 'type' may represent one or several species in tropical environments. These conceptual types become very useful when the wood anatomy does not allow for an unambiguous botanical determination.

3.8. Seed analysis

Seed and fruit remains from Leopard Cave and Geduld were separated from the surrounding sediment and charcoal matrix with the help of a binocular microscope. The same microscope, allowing magnifications of x8-x50, was used for observing and describing the morphological criteria (size, shape, ornamentation of the seed coat/ endocarp, etc.) that were used for the identification of remains. The botanical determination of seeds/fruits was more straightforward than that of charred wood remains. Many of the species were already well known by the author, and to confirm their identifications different sources were consulted. In the laboratory, *The Digital Atlas of Economic Plants in Archaeology* (Neef et al. 2011) as well as reference collections, in particular, those collected by the author in the field (see below) were referred to. Further confirmation could then be obtained by both Dr C. Sievers, a specialist in Southern African archaeobotany at the University of the Witwatersrand and by our informants within the Damara and San communities in the Erongo region (see chapter 4). Finally, comparisons with reference material held in the Botswana Botanical Garden, Gaborone, were also made.

3.9. Constitution of a reference collection

Given the difficulties mentioned above and the fact that Southern Africa has so far been the subject of few archaeobotanical investigations it was decided from the beginning to put effort into creating collections of wood and seeds/fruits representing the flora present in the study region. Not only is the reference collection crucial for the identifications and classification of the archaeological material for this study, but the aim is to set a foundation to build a larger regional catalogue for southern African reference collections for future use. Reference specimens were collected during the months of June and July in 2014, 2015 and 2016 from three different areas (Figure 39). On the one hand, samples were collected from the plant formations surrounding the sites of Toteng and Leopard Cave. More samples representative of southeastern Botswana were collected from the Botswana Botanical Garden (under the Botswana National Museum). Mr Mafokate from the Botanical Garden was particularly helpful in confirming the identification of the specimens through the examination of the herbarium sheets associated with the wood samples. For the identification of plants in the field, several books on the flora of Southern Africa, among them van Wyk and van Wyk (2007) and Coates-Palgrave (2002) were consulted.

Despite the distance between the three areas, they experience similar environmental conditions and partly host the same types of plant formations. Thus, when a particular species was present in quite a few of the chosen localities, for saving time, the species would be collected only once.

Woody parts in the form of tree trunks and parts of branches were cut with the aid of a cutter, an axe, and shears before being dried in the sun (Figure 40). In order to prepare the specimens for direct comparison with charcoal samples under a reflected-light microscope, they were burnt in a

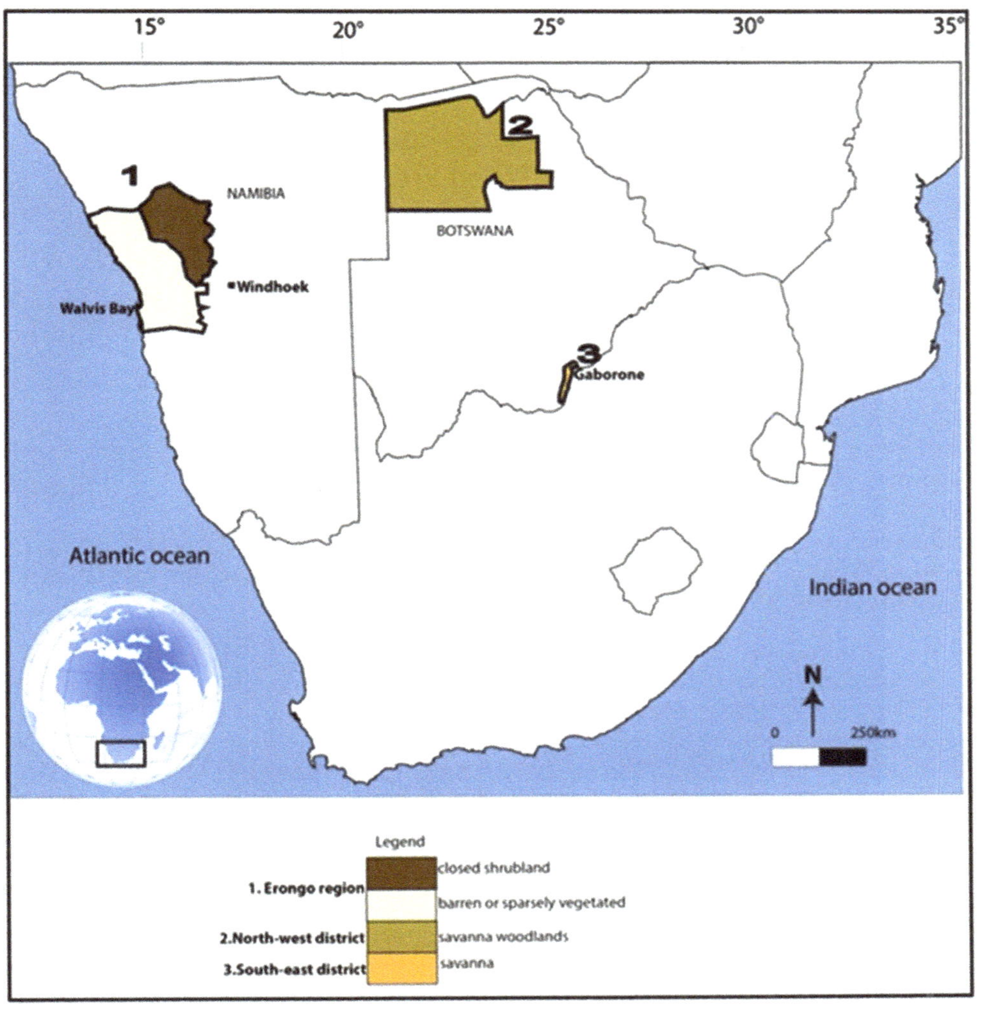

Figure 39. Areas in Namibia and Botswana where the reference material was collected. Map author: M. Mvimi.

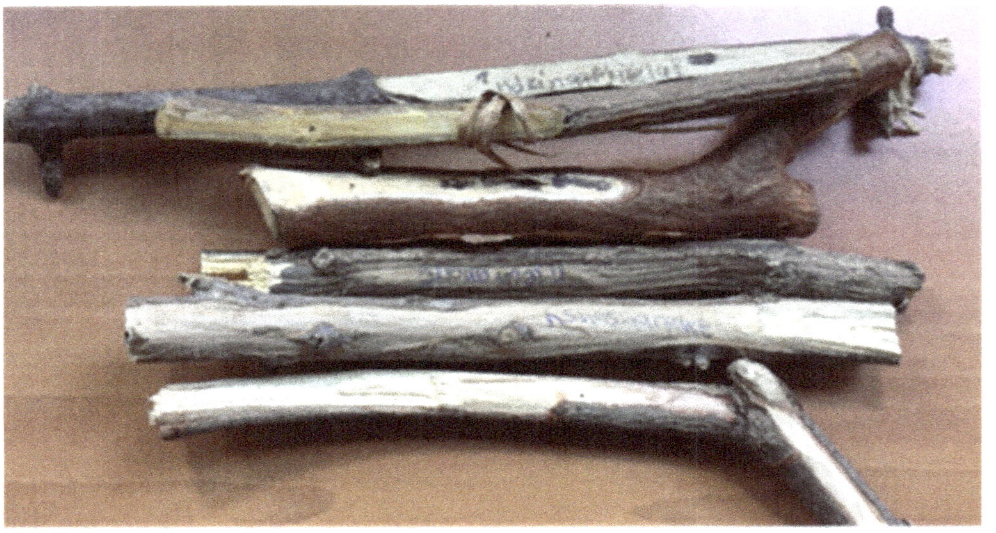

Figure 40. Labelled dried wood cuttings before charring. Photo credit: M. Mvimi.

muffle furnace at the National Museum of Natural History in Paris, France. Descriptions of the features were made following the criteria of the IAWA List of Microscopic Features for Hardwood Identification. Pictures were taken with a Scanning Electron Microscope (SEM), observations were made, and the results are given in the form of a catalogue where 22 specimens are described (see the appendices).

Other parts of the plant including seeds, fruit, inflorescence and sometimes roots were also collected. Leaves, flowers and other diagnostic elements were dry-pressed, labelled and organised in the form of a herbarium (Figure 41).

The constituted reference collection described above comprises a total of 49 species (Table 14) representative of 20 different families of the Angiosperm group (flowering plants). Among these 38 have provided a wood sample; 17 species have provided both a wood sample and other parts of the plants. Some families are better represented than others, for example, the Fabaceae (N=15) and the Combretaceae (N=5), because of the abundance of several species in the study areas.

Figure 41. An example of a herbarium sheet with leaves and fruits/seeds from Vachellia erubescens. Photo credit: M. Mvimi.

Table 14. List of species in the reference collection with their family, Latin and vernacular name as well as the part of the plant collected (F/INFL – Flower/inflorescence; FR/S – Fruit/seed)

N°	Family	Scientific name	Common name	Wood	F/INFL	Leaf	FR/S
1	Anacardiaceae	*Lannea schweinfurthii*	False marula	x		x	
2		*Sclerocarya birrea*	Marula	x			
3		*Ozoroa paniculosa*	Resin tree				x
4	Apiaceae	*Steganotaenia araliacea*	Carrot tree	x			
5	Bombacaceae	*Adansonia digitata*	Baobab	x		x	
6	Burseraceae	*Commiphora glandulosa*	Tall fire-thorn corkwood	x			
7		*Commiphora mossambicensis*	Pepper-leaved commiphora	x		x	

N°	Family	Scientific name	Common name	Wood	F/INFL	Leaf	FR/S
8		*Commiphora viminea*	Zebra-bark corkwood	x			
9	Capparaceae	*Boscia albitrunca*	Shepherd's tree	x			
10		*Boscia foetida*	Stinky shepherd's tree	x		x	
11	Combretaceae	*Combretum apiculatum*	Red bushwillow	x			
12		*Combretum imberbe*	Leadwood	x		x	x
13		*Combretum zeyheri*	Large-fruited bushwillow				x
14		*Terminalia prunioides*	Purple-pod terminalia	x			x
15		*Terminalia sericea*	Silver terminalia	x			
16	Euphorbiaceae	*Croton gratissimus*	Carrot tree	x			
17		*Bridelia mollis*	Velvet bridelia			x	
18	Fabaceae	*Albizia anthelmintica*	Worm-bark false thorn				x
19		*Vachellia erubescens*	Blue thorn acacia	x	x	x	x
20		*Vachelia erioloba*	Camelthorn acacia				x
21		*Senegalia mellifera*	Black thorn acacia	x		x	x
22		*Vachellia karoo*	Sweet thorn acacia	x			
23		*Senegalia nigrescens*	Knobthorn acacia	x			
24		*Vachelia robusta*	Riverthorn acacia	x			
25		*Vachelia tortilis*	Umbrella thorn	x		x	x
26		*Adenolobus garipensis*	Butterfly leaf	x			
27		*Cassia abbreviata*	Long-tail cassia	x		x	
28		*Colophospermum mopane*	Mopane	x		x	
29		*Dichrostachys cinerea*	Sickle bush	x	x	x	x
30		*Mundulea sericea*	Cork bush				x
31		*Ormocarpum trichocarpum*	Caterpillar pod			x	
32		*Peltophorum africanum*	African wattle			x	x
33	Loranthaceae	*Tapinanthus oleifolius*	Mistletoe			x	x
34	Malvaceae	*Azanza garckeana*	Snot apple	x		x	
35		*Dombeya rotundifolia*	South African wild pear	x			
36	Moraceae	*Ficus abutilifolia*	Large-leafed rock fig	x			
37	Moringaceae	*Moringa ovalifolia*	Motinga	x			
38	Olacaceae	*Ximenia americana*	Blue sourplum	x		x	
39		*Ximenia caffra*	Sourplum	x			
40	Rhamnaceae	*Berchemia discolour*					x
		Ziziphus mucronata	Buffalo thorn			x	x
41	Rubiaceae	*Gardenia volkensii*	Bushveld gardenia	x		x	
42		*Vangueria infausta*	Wild medlar	x			
43	Sapotaceae	*Sideroxylon inerme*	White milkwood	x			
44	Sterculiaceae	*Sterculia africana*		x			
45	Thymelaeaceae	*Corchorus olitorius*	Bush okra		x	x	
46	Tiliaceae	*Grewia flava*	Velvet raisin	x	x	x	x
47		*Grewia flavescens*	Sandpaper raisin	x			
48	Vitaceae	*Cissus cornifolia*	Ivy-grape	x		x	x
49		*Cyphostemma currori*	Cobas	x			

4

Plant Use Among Present-day Damara and San Populations in the Erongo Region, Namibia

Contact may take many forms, not all of which lead to dependency, abandonment of foraging or incorporation into "more powerful" social formations (Solway and Lee 1990)

NB: All the names of the participants were modified

4.1. Objectives of the ethnobotanical study

Reconstructing past practices from the archaeological remains of material culture can be quite challenging. The record at hand is often insufficient and patchy, more so in archaeobotany where the bulk of organic material has disappeared through time. Moreover, the lack of comparative work from the study area adds to the difficulty of understanding past environments and subsistence activities.

Even though there is no direct comparison between the Late Stone Age communities and today's populations in the Kgalagadi basin today and, ethnobotanical enquiries can provide hints on the exploitation and the use of local plant resources, some of which may have been present in the region in the past. Some researchers dispute this approach, but it is helpful when it comes to interpreting material from archaeological contexts. It can put people into spaces with objects they move with and manipulate, allowing us to envisage these actions even in contexts from which the actors have since long disappeared.

Understanding and reconstructing the history of the people and their culture as it applies to archaeological inferences, requires letting the populations under study to participate actively in the investigation. Current ethnographic work seems to be heading in this direction. Recently the providers of traditional ecological knowledge (TEK) are seen increasingly at the centre of these studies. They help researchers avoid depending on pre-fabricated fixed models when interpreting archaeological data. Omitting their knowledge can result in the inclusion of redundant points of views mired in fixed classifications. In the region, a good example of TEK inclusion is the application of ethnographic studies that incorporate linguistic and archaeological questions (Rapold and Widlok 2008).

In order to gather general information on the exploitation and use of plant resources in the arid environment of the Erongo region, an ethnobotanical survey was conducted within two different communities living close to the site of Leopard Cave: a group of sedentary inhabitants from the Damara ethnic group (Khoekhoegowab (Nama/Damara) living in the small village of Tubusis and members of the Ju/'Hoansi the San population demonstrating traditional plant uses in the framework of the San Living Museum, Grashoek. Surveys took place in September 2017 and October 2018. Ms M. Mwantondange from the University of Namibia took part in the ethnobotanical surveys. Questions were oriented towards issues of special interest for our understanding of past plant uses such as the collection of wood for fuel and construction, the role of wild plants in the diet, knowledge and use of medicinal plants, seasonality of foraging, etc.

4.2. Description of the Damara village of Tubusis

The small village of Tubusis (21° 32′ 2″ South, 15° 27′ 4″ East) is located on the left side of the gravel road D306 leading to Usakos, on the western foothill of the granitic Erongo Mountain and within the Gaingu Communal Conservancy (Figure 42) (Angula, 2010). The village is situated about 60 km from Usakos, 50 km from Spitzkoppe, 80 km from Okombahe and Omatjete and 60 km from Omaruru (MTC Namibia EIA report 2018).

The local climate is semi-arid to arid with mean annual temperatures of 18°C and peaks reaching around 40°C during the hot season. The mean annual rainfall, mostly occurring as summer precipitation, varies between 100 and 500 mm. The arid conditions influence the vegetation cover that belongs to the desert/Nama Karoo biome, also known as the "Karoo shrubland" of northwestern Namibia. It is characterised by open to very open scrublands dominated by drought-resistant species from the *Fabaceae* (*Mimosoideae*) and *Capparaceae* families as well as seasonal grasses and other herbaceous plants, among them several creepers (Mitchell 2002, Sletten et al. 2013).

As of 2011, around 800 persons were resident in Tubusis (2011 Namibia census). Most of the middle-aged, working-class inhabitants live and work in nearby towns such as Swakopmund, Usakos and Omaruru and as far as Windhoek. Usually, they return to the village to spend time with their families on weekends and public holidays. Retired and unemployed adults, as well as small children and those going to primary school (6–14-year-old), reside permanently in the village. Grandparents look after the young children in the absence of their working parents. The great majority of the population belongs to the Damara/Nama ethnic group with a few exceptions corresponding

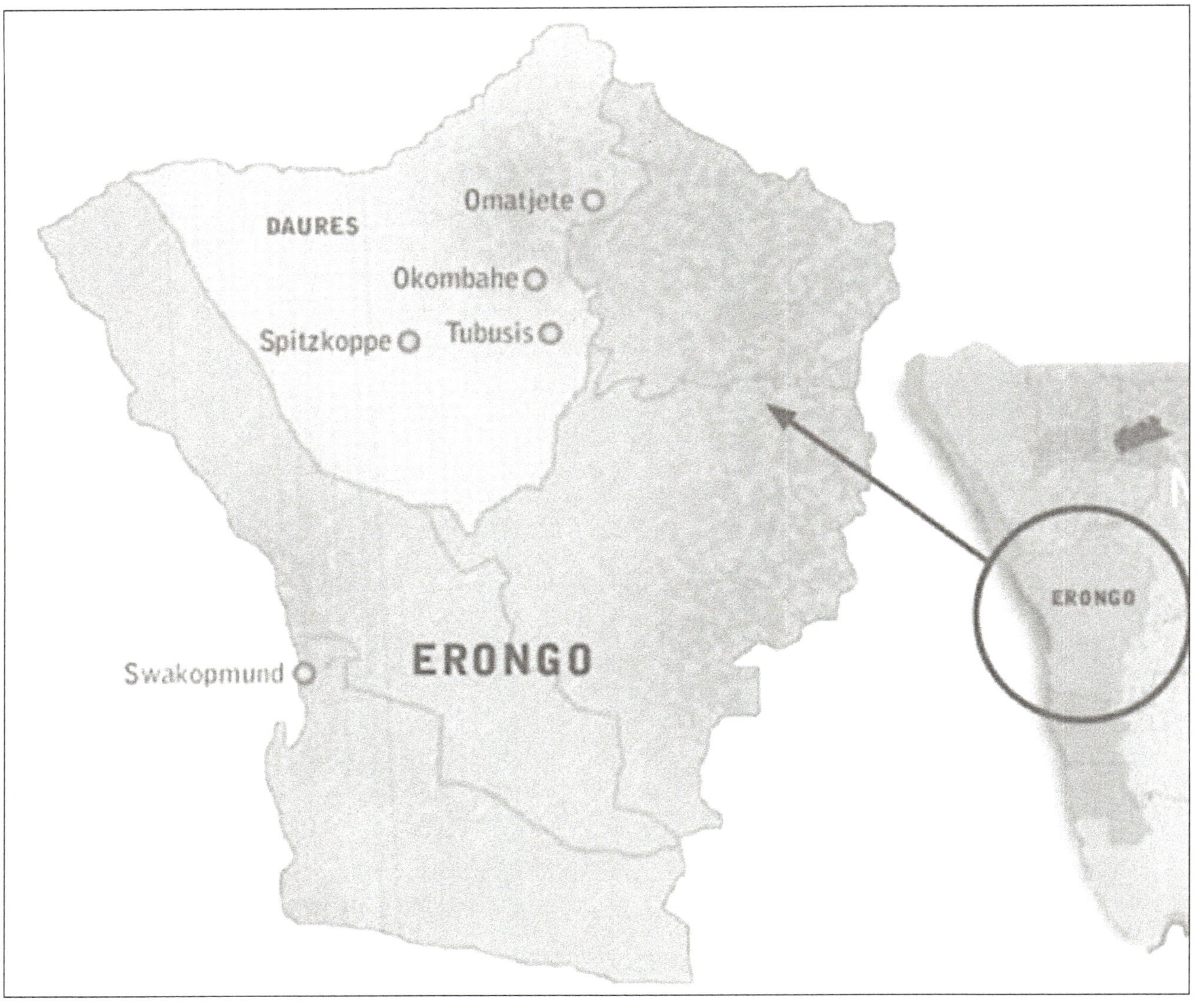

Figure 42. Location of Tubusis village in the Erongo region (after Angula, 2010).

to persons from other areas who arrived through marriages or work or who run commercial activities.

In addition to around 130 homesteads, the village comprises a primary boarding school, a church, two supermarkets, a communal water-tap and two drinking spots. For amenities, including groceries and medical care, the inhabitants commute to the nearby towns of Usakos and Omaruru. A gravel road runs through the settlement where commuters hitchhike to reach the nearby towns. In spite of the village's proximity to larger and more modern towns, its inhabitants still acquire the majority of their resources for everyday needs from the natural environment, mostly because of their modest social and financial status.

The opposite side of the road is used as grazing ground and for collecting fruit, sap or small prey such as tortoise, rodents and birds. Photos of this area (Figure 43, Figure 44) were taken in October 2017 i.e during the late dry season towards the beginning of the rainy season. There is almost no ground cover and most of the trees, except the evergreen *Boscia* species, have lost their leaves.

Figure 43. View of part of Tubusis village and it's surroundings. Photo credit: M. Mvimi.

Barren as it may seem, most of the vegetal material used in the village still comes from this type of biotope.

4.2.1. The Damara ethnic group

The Damara are a Khoekhoegowab dialect speaking group found in Southern Africa making 8.5% (about 100 000 inhabitants) of Namibia's present-day population

Figure 44. Landscape opposite Tubusis village with granite foothills in the background and open *Vachellia erubescens* woodland with a grazing herd of goats. Photo credit: M. Mvimi.

(H. Soodyall and Jenkins, 1993). One of the oldest tribes in Namibia, the Damara speak a Khoisan language, but they have a different biological history altogether. While they are currently found in the north-central parts of Namibia in the Kunene region, some parts of the Erongo region and also spread in other parts of Namibia (Figure 45) mostly for professional activities (Mendelsohn 2008) the origin of this group is still poorly understood. A number of studies on this subject are however currently underway. The Damara were previously classified as Khoehkoen and the indigenous and earliest inhabitants of the area that is now Namibia, but linguistic studies reveal that they speak the Damara Khoekhoegowab dialect language that many scholarly writings ascertain they inherited from their Nama slave masters. The study of mitochondrial DNA suggests that they are closely related to the Herero Bantu group. On the other hand, lexical comparison between the Central Khoe and Damara Khoekhoegowab groups give a glimpse of their relationship, possibly raising questions on why a Khoe language and a Negroid non-Khoe or San group would share language aspects, but it cannot give answers about the origins of the Damara people (Haacke, 2008). One school of thought suggested they could be one of the first inhabitants of the region to be subordinated into servitude by their later arrivals with livestock. Most Bantu herd goats and sheep together and the Damaras are no exception; today they are goat and sheep herders and some of these conform to the names "Damara goat" and "Damara sheep" (Pleurdeau et al., 2012) which are well adapted to harsh and arid environmental conditions (Badenhorst, 2006). Sheep though are much earlier domesticates for the Khoekhoen and they give the perception that the Khoekhoen may have originally favoured sheep over goats Apart from livestock farming and the infertile land for ploughing, the Damara also rely highly on wild seeds, fruits, honey and grass grains for subsistence (Sullivan, 1999)

4.2.2. Methodology

After walking around the entire village of Tubusis (about 800 m from one end to the other), five homesteads were sampled in order to methodically interview their inhabitants and gather information on their knowledge of local plants and their uses. The informants were mainly selected according to the villagers' recommendations citing the presence of a particularly knowledgeable (often an elderly) person (Figure 46).

The interview method entailed the conducting of open-end questions with responses noted down while a voice recorder recorded the whole interview process. Participatory observations were also made, and some informants led "guided visits" in the area. Overall, these provided detailed direct descriptions of different households, their surroundings, vegetation cover and the use of plants. Photographs were taken and videos recorded

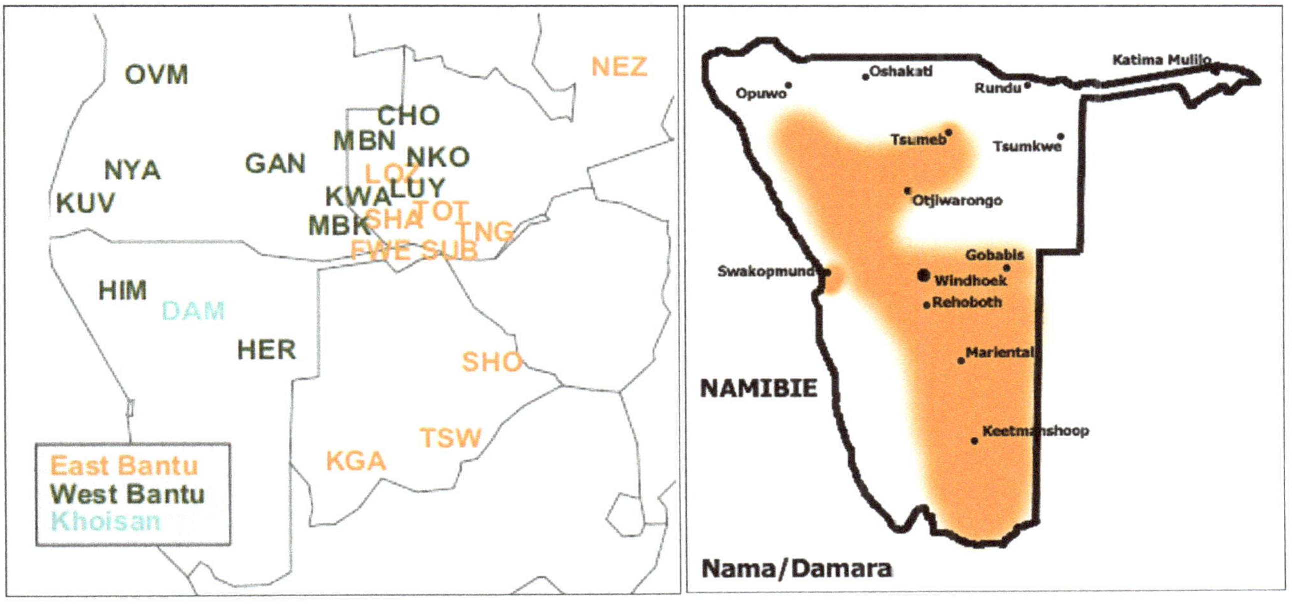

Figure 45. Map of Namibia showing the current distribution of Damara populations; orange colour on the left and blue"DAM" on the right (Sources: (Wikipedia 2005; Barbieri et al., 2014).

Figure 46. A Damara woman and man living in Tubusis. Photo credit: M. Mvimi.

in necessary instances as support material. Later, all the above media were transcribed.

Concerning the plant taxa, the majority of the names were indicated in the Khoekhoegowab language. Both the vernacular and Latin nomenclature is applied in this chapter.

Below is a description of the information obtained from each of the chosen households. As can be noticed, the information on plant use varies according to the composition of the household and family practices. At the end of this chapter is a summary table comparing the uses of plants by the two groups living in the same type of environment at a few kilometres distance.

Because the visit to Tubusis took place during the dry season, not all plant uses could be directly observed. For example, many fruits were not available in October, and only a selection of vegetative parts of these species could thus be observed.

4.2.3. Description of plant uses for each household

4.2.3.1. Household 1

The head (matriarch) of this household is Mrs Emma Justus who has lived in Tubusis since 1987 and works in a hostel. Her granddaughters Emma Justus Jr., 10 and Nelly Gaubis, 25 were present during the visit.

As one of the village elders, Mrs Justus is regarded as very knowledgeable about the history of the village and about plant use. She was particularly conversant on medicinal plants that she harvests, prepares and administers to her patients. She also provided information on the use of plant materials in the architecture of her house, plant foods and the surrounding vegetation. She made demonstrations using living plants and referred to others not directly observable in the immediate surroundings.

The fence surrounding Mrs Justus's house and its yard have been constructed by associating standing gum-poles (store-bought treated *Eucalyptus*) and locally collected stems and branches. Among the latter, we could identify *Vachellia erubescens* and *Senegalia nigrescens* that accounted each for around one-third of the fence material. *Boscia albitrunca* was also used (around 1/5) as well as some branches from species that were not identifiable. To maintain the upright position of the palisade, poles were dug into the ground and spaced at an average distance of 1.5 m and a horizontal wire was used to keep them together.

Mrs Justus's late husband built the house from scratch, though she does not recall the exact year of construction. Several square or rectangular structures made from mud, concrete bricks, corrugated iron and provisional material like garden mesh and empty plastic bags surround a rectangular yard measuring approximately 20 m × 30 m. The largest of these structures is used as the bedroom by the whole family and the courtyard in front of it as the sitting and cooking area. A second structure located at the far corner of the yard is used as storage for firewood and kitchen utensils and can also be converted to a living area in the rainy season. Mrs Justus mentioned no specific choice of wood for construction and combustion but indicated that wood is mainly acquired according to its availability and sometimes durability. She also attested that firewood, which is only collected as dry wood, has recently become scarce around Tubusis and is therefore brought in from neighbouring villages and farms. Building material is obtained from felled trees, usually in the proximate surroundings of the village but at the present time, wood used for construction is also collected from further afield in order to avoid exhausting fruit resources, general food supply and animal feed from sparsely vegetated areas. She further points out that *Vachellia erubescens* is a tree she deems exceptionally essential as it produces resin that she and her grandchildren collect to enjoy as a delicacy or for selling. She underlines that this collection is occasional and that is the reason why the resin is not for everyday use. Moreover, Mrs Justus provided specific information on the uses of fruit and wood from *Spirostachys africana*, *Boscia albitrunca* and *Vachellia karoo*.

4.2.3.2. Household 2

The main informant in the second household, visited on September 6th, 2017, was 34-year-old Mrs Meda !Kharuxab. Six other family members were present at the time of interview. Mrs !Kharuxab interchanges residences between Tubusis and her farm, situated 10 kilometres away, where her husband permanently resides. She was born and raised in Tubusis by her grandmother who was the village's traditional doctor. She attests that her knowledge of plant use both as medicine and for everyday use was instilled in her by her grandmother.

When in Tubusis, Mrs !Kharuxab lives with the six members of her extended family consisting of nieces and cousins. The homestead they live in used to belong to her late grandmother. Just like in household 1 this yard is surrounded by a fence made with a mixture of eucalyptus poles, one half *Boscia albitrunca* and one fourth *Vachellia erubescens* completed by a mixture of other unidentified wood elements. The compound encompasses three sheds made from an assortment of corrugated iron and wood, including that of different *Vachellia/Senegalia* species. Two of the structures act as bedrooms and the third and smallest of the three is the kitchen and storage area for both firewood and foodstuffs. Some firewood is however also stored outdoors adjacent to the kitchen. Cooking takes place outdoors the majority of the time and inside the kitchen during the rainy season. A *!nxara* tree stands by the compound entrance and the family sit under its shade during the day to cool off. Cacti and sisal plants were planted in the backyard for their medicinal virtues by family members. Mrs !Kharuxab's interview included a tour to explore plant life in the surrounding areas in the unpopulated space opposite the households and the mountainous area behind the village. Consequently, from this household, a higher number of plant species including *Boscia albitrunca*, *Vachellia erubescens*, *Commiphora wildii* (Figure 47), *Lippia javanica*, *Aloe gariepensis*, *Harpagophytum procumbens* and *Terminalia prunioides* among others were covered.

4.2.3.3. Household 3

This household comprised only one person: Mr Solomon ||Hoëb, 42 years old, who is a teacher. Mr ||Hoëb is originally from Tsumkwe but he has been living in Tubusis for the past ten years where he teaches at the local primary school. He is a tenant of a one bed-roomed brick house. The rectangular yard is fenced with a mixture of palisades from different tree species that are put together with galvanised wire. The palisades are relatively old and worn out therefore not very easy to identify. Since he was not present at the time of construction, the respondent cannot be relied on to give the wood types used for building but states that to the best of his knowledge, wood is usually chosen on the basis of durability and resistance to parasites. He however commented on the plants found around his homestead and the school area. He knows especially well some species that he uses on a daily basis: *Vachellia karoo* and *Vachelia tortilis* for fuel and *Boscia albitrunca* and *Boscia foetida* for fruit. Some of the wood from these trees are found by his fireplace nearby the house but mentions that he uses a gas stove to cook most of the time.

Figure 47. Mrs Meda !Kharuxab demonstrates the use of *Commiphora wildii* while being recorded on a voice recorder. Photo credit: M. Mvimi.

4.2.3.4. Household 4

The 60-year old woman who was the fourth informant preferred to remain anonymous. She stays at home and tends to her goats, chicken and sheep. She has grown children who are working in different towns in Namibia. The fence of her house consists of one-fifth gum poles associated with local branches from one fifth *Boscia albitrunca*, one-fourth *Vachellia/Senegalia* spp. as well as other unidentified plants. Her homestead measures 30 m × 30 m and the two-bedroomed house is located at the back leaving a large courtyard in the front yard where firewood is stored and cooking takes place. A *Vachelia tortilis* tree growing near the entrance serves as sitting and rest space during the day. Besides stating the use of *Boscia albitrunca*, *Boscia foetida*, *Ziziphus mucronata* (food) and *Vachelia tortilis* (fodder) the informant pointed out the seeds of a grass species (genus *Sporobolus*) that is collected from ant mounds and eaten raw (Figure 48).

4.2.3.5. Household 5

Mrs Esther Ganuses, 72 years old, lives in one house together with three of her relatives, two females and one male. She is one of the first inhabitants of Tubusis but does not recall the year she first settled there. She is the current traditional healer of the village. Her homestead is built of material similar to that observed in the previously investigated households (Figure 49), namely, a mixture of wild woody plants, mesh and galvanised wire (Figure 50). The walls and roof of her two houses are made with corrugated iron, and the kitchen house is made up of mesh wire and corrugated iron roofing. Firewood and cooking utensils are kept in this area. Excess wood and abandoned kitchen utensils are kept behind this structure. The rooftop is used as a scrap area. As the village traditional healer, the informant has a very wide knowledge of medicinal plants. She selects carefully the wood that she uses for fire, unlike most of the other informants. She

Figure 48. Edible *Sporobolus sp.*, grass seeds that are collected from around ants' nests. Photo: M. Mvimi.

Figure 49. Fencing around the yard of Mrs Esther Ganuses. Note the mixture of metal sheets and tree branches. Photo credit: M. Mvimi.

Figure 50. Wood used for constructions and fuel stored beside the kitchen house in household 5. Photo credit: M. Mvimi.

believes that certain species are taboo, maintaining that they are capable of bringing bad fortune to the household in which they are used and that the bad luck is easily propagated by the smoke from the fire. Mrs Ganuses also demonstrated how dried *Ziziphus mucronata* and *Combretum imberbe* leaves are used for treating coughs.

4.2.4. General observations

Observations together with the information provided by the informants reveal an elevated exploitation of almost all of the locally available plants for consumption (food, fuel and medicine), construction and commercialisation (Table 15). For the village architecture (fences and houses), the inhabitants use a combination of traditional and modern building materials. Most house walls are constructed using cement bricks, corrugated iron sheets, branches and various textiles while some combine all the aforementioned materials, merged with wood. The yard fences are also made with a combination of barbed wire, galvanised wire, wood and treated gum poles. Most of the residents have either constructed the structures themselves or received help from relatives to do so.

Free-range pastoralism is practised on a relatively small scale by keeping drought-resilient livestock such as goats, sheep and poultry. Pet cats and dogs are also common. Not much large stock like cattle are kept but donkeys and horses are used as modes of transportation and for drawing carts. During the dry season, when grazing land is exhausted, the animals browse on evergreen trees like *F. albida* or *Boscia* species and the owners usually aid their livestock by collecting and storing (in a dry state) wild fruits mostly wild melon (*Citrullus caffer*, *C. ecirrhosus*). Animal feed is also attained from various trees from the Mimosoideae (Fabaceae family), such as *Vachellia/Senegalia* and *Dichrostachys cinerea,* by way of shaking the trees to make their nutritious pods fall to the ground. Since the family members residing permanently in the village are usually without formal employment, they alternate in accompanying the animals to feed in the open wood and scrublands during the day.

Almost no cultivation is practised by the villagers because of inadequate rainfall. The available village water supply is for domestic use and leaves no surplus for irrigation purposes. The villagers interviewed mentioned farms situated at a distance of about 7–10 km where herds of cattle are kept and commercial vegetables such as cabbage, carrots, beetroot etc. are cultivated, employing a substantial number of Tubusis residents. For irrigation and watering livestock, these farms use boreholes.

Dry wood for fuel is collected from the woodlands and is stored in the courtyard or in a roofed space during the rainy season. When collected at distances further than 5 km, a donkey or horse-drawn cart is generally used to transport the wood. In 85 percent of the households, fire is made in a sheltered hearth in the courtyard or, when the weather is bad, inside the kitchen area. Usually, all family members are involved in the collection of wood but in some cases, wood loads are bought from vendors who bring it to the village in horse-drawn carts.

4.3. The San Living Museum in the Omandumba farm

The second ethnic group living near Leopard Cave concerned with the ethnobotanical survey are the San. Members of the Ju/'Hoansi local community demonstrate traditional activities in the "San Living Museum" founded in 2008 at the Omandumba farm, Grashoek. For the purpose of this study, this open-air museum was visited on two occasions in the months of September 2017 and 2018 (Figure 51). During the day various activities relative to the San way of living are organised, and the information gathered for this study was acquired during several guided tours. The demonstration took place in the San settlement and then in the surrounding landscape. The residence entails a makeshift compound, cleared of groundcover. The community does not live in permanent houses but construct temporary huts made of grass and/or plastic bags. The "workplace" is under a boulder and the temporary homestead is behind this spot. Throughout the day, each member of the family performs tasks primarily for tourist attraction. The women do craftwork like beading; manufacturing ornaments like necklaces, bracelets and decorations while the men do the guided tours, gather the material needed to make the crafts and several other types of crafts like wood-carving. The children mostly assist the adults but at the same time, they play around and take care

Table 15. Summary table of the uses of woody plants in the Tubusis Damara community

Family	Scientific name	Common name	Damara name	Uses	Presence in the archaeobotanical record
Agavaceae	*Agave sisalana* Perrine.	wild sisal	• Damara name unknown	• Leaves are crushed and mixed with other herbs for various medicinal uses : immune booster, pain killer, anti-inflammatory, snake repellent.	absent
Burseraceae	Commiphora wildii Merxm	Namibian Myrrh	andosia	• Both wet and dry leaves are used as a perfume or deodorant. Gonimbrasia belina host; the sap is a general anti-inflammatory.	possibly present
Capparaceae	*Boscia albitrunca* (Burch.) Gilg & Gilg-Ben.	shepherd tree	ǁhüni.s	• Wood is collected dry for firewood; the bark is chewed dry; as a medicine the dry bark is chewed dry, in powder form or boiled and used to treat diahorrhea or general ailments; the root is added to sour milk as a thickening agent; fruit is eaten fresh or soaked overnight to make juice.	charcoal and seeds
Capparaceae	*Boscia foetida* Schinz.	smelly shepherd tree	xắübȅ	• Dry wood is collected and used as fuel;leaves are chewed to treat stomach ailments or insomnia or soaked in water to ward off evil spirits; the evergreen tree provides shade.	charcoal and seeds
Celastraceae	Gymnosporia senegalensis (Lam.) Loes	confetti tree	!xhoes	• The dry wood is used as firewood; the leaves are soaked in water to cool it.	absent
Combretaceae	*Terminalia prunioides* M.A. Lawson	purple pod terminalia	ǂKhēas	* The leaves are used as animal fodder; the colourful pods are used as a decoration.	charcoal and seeds
Cucurbitaceae	Citrullus lanatus (Thunb.) Matsum. & Nakai	tsamma melon	tsama.s	the melons are used as animal feed, usually eaten in the wild by the animals; the seeds are rtoasted and eaten as a snack.	seeds
Euphorbiaceae	*Spirostachys africana* Sond.	tamboti	tamboti	• Wood is used as building material	absent
Fabacaeae-Mimosoideae	*Acacia erubescens* Welw. ex Oliv.	blue thorn acacia	!xhuung	• Wood is used as firewood and building materials especially uesd for fencing;the sap is edible and is sold around the village of Tubusis; the dry pods are used as animal feeds; powder from the stem, generated by boring insects is applied to the skin as a beauty product.	possibly present
Fabacaeae-Mimosoideae	*Acacia erioloba* E.Mey.	camel thorn acacia	ǀgáná.b	• Dry wood is used as firewood and the stem is used in fencing; the bark is kept indoors as an freshner; the leaves are used as fodder for animals	possibly present
Fabacaeae-Mimosoideae	Faidherbia albida (Delile) A.Chev.	apple-ring acacia	anas!	• Evergreen leaves are fed to animals	possibly present
Fabacaeae-Mimosoideae	*Dichrostachys cinerea* (L.) Wight & Arn.	sicklebush	ǀgöé.s	• Wood is used as firewood, usually when firewood is rare	charcoal
Fabacaeae-Mimosoideae	*Vachellia/Acacia karroo* (Hayne) Banfi & Glasso	sweet thorn acacia	ǁkhùű.s	• Dry wood is used as firewood; the fruit is eaten fresh, just before it reaches maturity and the dried fruit is used as animal feed.	possibly present

(*Continued*)

Table 15. Summary table of the uses of woody plants in the Tubusis Damara community (*Continued*)

Family	Scientific name	Common name	Damara name	Uses	Presence in the archaeobotanical record
Pedaliaceae	*Harpagophytum procumbens* Burch.	devil's claw	gamagu	• The root treats all illnesses; anti-inflammatory or ''blood cleanser''	absent
Poaceae	Sporobolus spp.	possibly giant rat's tail grass	sangwee	• The seeds are harvested from ant's nests for human consumption	possibly present
Rhamnaceae	*Ziziphus mucronata Willd*	buffalo thorn	x!aros	• The root is crushed and used against infections during childbirth	charcoal and seeds
Sterculiaceae	*Sterculia africana* (Lour.) Fiori	tick tree	aarup	• The bark is crushed and taken orally in powder form; fresh sap is used against gastric problems; the leaves are also medicinal	absent
Verbanaceae	*Lippia javanica (Burm.f.)* Spreng	fever tea/lemon bush	ho!xhobrosi	• The leaves are boiled and taken as an infusion	absent
Asphodelaceae	*Aloe gariepensis Pillans*	gariep aloe	aloe	• The leaves are crushed as are used as a general anti	absent
Cactaceae	*Cylindropuntia fulgida* Engelm.	boxing-glove cactus	• *referred to as cactus*	• The leaves produce a bitter sap that is used to treat high blood pressure, as an anti-inflammatory and as an immune booster; it is highly sought after and it is a protected plant in Namibia	absent

Figure 51. Exchanging with the informants at the San living Museum. Photo credit: A. Nankela.

Figure 52. The landscape in Omandumba West adjacent the San Living Museum. Photo credit: M. Mvimi.

of their younger siblings. During the tours, at least three informants are designated; amonst them, one must speak a Ju/'Hoansi dialect and another to translate into English.

The landscape around the Omandumba farm (Figure 52) is very similar to that of the surroundings of Tubusis with only a slight difference in vegetation cover. Thorny *Vachellia/Senegalias* and grasses dominate the open scrub/woodlands but while *Ziziphus mucronata* is present in Tubusis, *Grewia* and *Ximenia* species are the most common fruit trees in the San area. *Boscia albituruncа* and *B. foetida* grow in both places and the fruits of these trees are eaten raw or fermented.

4.3.1. The San populations of Southern Africa

Designated as *Bosjemans* or *Bushmen* by the colonisers, the aboriginal communities of hunter-gatherers living in the Kgalagadi desert of Southern Africa are at present preferably called San by academics, a name derived from the local terms *Sonqua* and *Soaqua* (Smith and Ouzman 2004). The San are today found in Namibia, South Africa, Botswana and Zimbabwe (Figure 53). 38 000 of them live in Namibia, making up around 2 percent of the country's population. Not all San people are related and they usually prefer to be referred to by their respective group name (Dan et al. 2010).

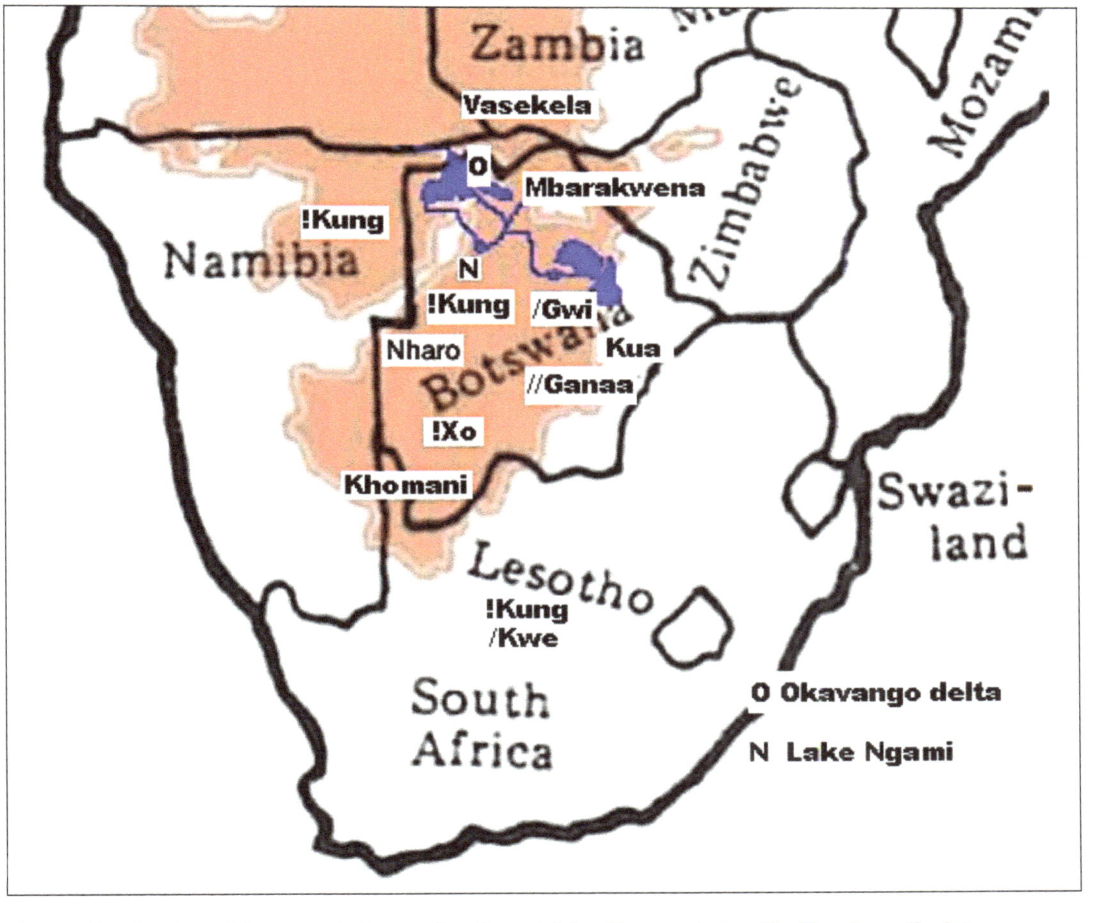

Figure 53. The distribution of San populations in Southern Africa (Source: https://laulima.hawaii.edu).

Unlike the Damara group, the lifestyle and origin of San populations "who do not possess pottery nor raise livestock" have attracted quite some attention since the 18th century. John Philip, superintendent of the London Missionary Society and a militant against the San genocide of the late 18th and early 19th centuries, was together with his colleagues the first to publish on the "Bushmen" in the Cape Province (South Africa). In 1907 the German photographer S. Passarge wrote and illustrated a book entitled *Die Buschmänner der Kalahari* in which he describes his encounters with the indigenous people of the Kgalagadi in present-day Botswana and Namibia (Demangeon 2018, Penn 2013). This description is qualified by Wilmsen et al. (1990) as "San research that came close to an organised anthropological study".

A "Great Kalahari San debate" is still open as to the origin and identity of the San populations (Kurtz 1994). Their linguistic as well as their genetic structure raises a number of questions as do their links to the Khokhoen ethnic groups that are also present in the region. Indeed, the two are sometimes designated under the common name of Khoisan. At times, the Khokhoen have been defined as "San populations with livestock" and the San as "Khokhoe groups without livestock". In the 1970s and 1980s, scholars proposed a model in which Khoekhoen herders migrated with sheep into territories previously occupied by the San. Still, there are contrasting theories on the relation between these groups and how one of them became herders (Barnard 2008). Did the ancestors of the Khoekhoen populations arrive in Southern Africa in the past 2000 years bringing with them livestock or did the Khoekhoen and San co-exist in the region as hunter-gatherers prior to the adoption of herding by the Khoekhoen as the result of interactions with Bantu populations? The quest to interpret the culture and identities of the San *versus* the Khoekhoen has been going on for almost five decades, with by far more scholars favouring the idea that some local groups moved from a hunting and foraging way of life to become herders (Smith and Ouzman 2004, Soodyall et al. 2008). Henn et al. (2008) further argue for a possible migration from Tanzania of the Sandawe people who share a Y chromosome trait with the Khoekhoen and San groups of Southern Africa. This migration occurred allegedly independent of the Bantu migration. Genetic studies show that San represents one of the oldest lineages known among modern humans (Schuster et al. 2010). The analysis of mitochondrial DNA further shows that the San and the Khoekhoen might share a common matrilineal ancestor while the Y chromosome shows a greater variation between the two groups (Soodyall et al. 2008). Certain phenotypic traits within the San group suggest adaptations to arid environments by means of storing fats in certain body tissues as discussed by Schuster et al. (2010).

While the genetic approach helps us to appreciate the relationship among different ethnic groups, the cultural and social dimensions of the evolution of aboriginal populations are more difficult to grasp mainly due to a lack of sufficient palaeoanthropological and archaeological data (Behar et al. 2008). One well-known dimension of the San material culture is the rock art found in Southern Africa in the form of engravings and paintings believed to have been produced by forager communities. Though most of their activities are depicted in rock art, possible interactions of these people with other groups calls for further revision of the rock art protocol, as it may be a representation of art derivatives from the groups they interacted with.

San and Bantu groups have been involved in ivory trading as early as the 17th century (first written record from 1661), and the Xhosa later traded ivory with the Portuguese. In the 19th century, there is evidence of intermarriages between the San and these Bantu groups (Jolly 1996). In Namibia, recent San encounters with other groups resulted in their population being reduced by ethnic killing as was the case with the German massacre of Herero, Damara and San communities in 1904, and displacements of the Ju'hoasi groups from their land into the Tsumkwe administrative centre (Figure 54). More displacements followed in the 1950s and 1960s resulting in most San converting to livestock herding (Barnard 2008, Dan et al. 2010).

4.3.2. Plant use in the San community from Tsumkwe

The San community participating in the activities of the San Living Museum at the Omandumba farm belongs to the San populations from the Tsumkwe region. During the first visit, the principal guide and informant was Tshaa aged 36 from Tsumkwe who has a good command of the English language. His counterpart, 33-year old Xhasa is also from Tsumkwe but only spoke his native language. !Nokhwe #Mxaa, 24 years, and Stefan Herman, 25 years, both born in Aasvoelnes in the Tsumkwe region, were the informants for the second leg of the mission undertaken in September 2018. Various plants play a significant role in the daily lives of the Ju/'Hoansi, from food to tool making and medicine (Table 16). Most of the tools are related to hunting and for processing other materials such as plants and animal hides. During the guided tours in 2017 and 2018, the informants described the characteristics and uses of more than 20 local plant species based on the know-how they had acquired from their parents and grandparents. Besides demonstrating plant uses on the basis of species currently growing around the Omandumba farm, the guides also explained the interactions with trees and herbs featured in the local rock art or plants that are not exploited by humans but consumed by wild animals. Furthermore, plants that were deemed useful to the San culture but absent from the Omandumba environs were also used in some of the demonstrations. An example is the Mongongo tree (*Schinziophyton rautanenii*). The tree occurs in a relatively narrow belt across the subtropical latitudes of Southern Africa and for several decades it has been a food source for indigenous communities, notably Khoisan groups (Mitchell, 2002a; L. H. Robbins et al., 2008) . The nut kernels contain protein, magnesium, iron, zinc, fats and vitamin E (Lee 1973). Moreover, branches of the tree are used to create pallets that play a pivotal

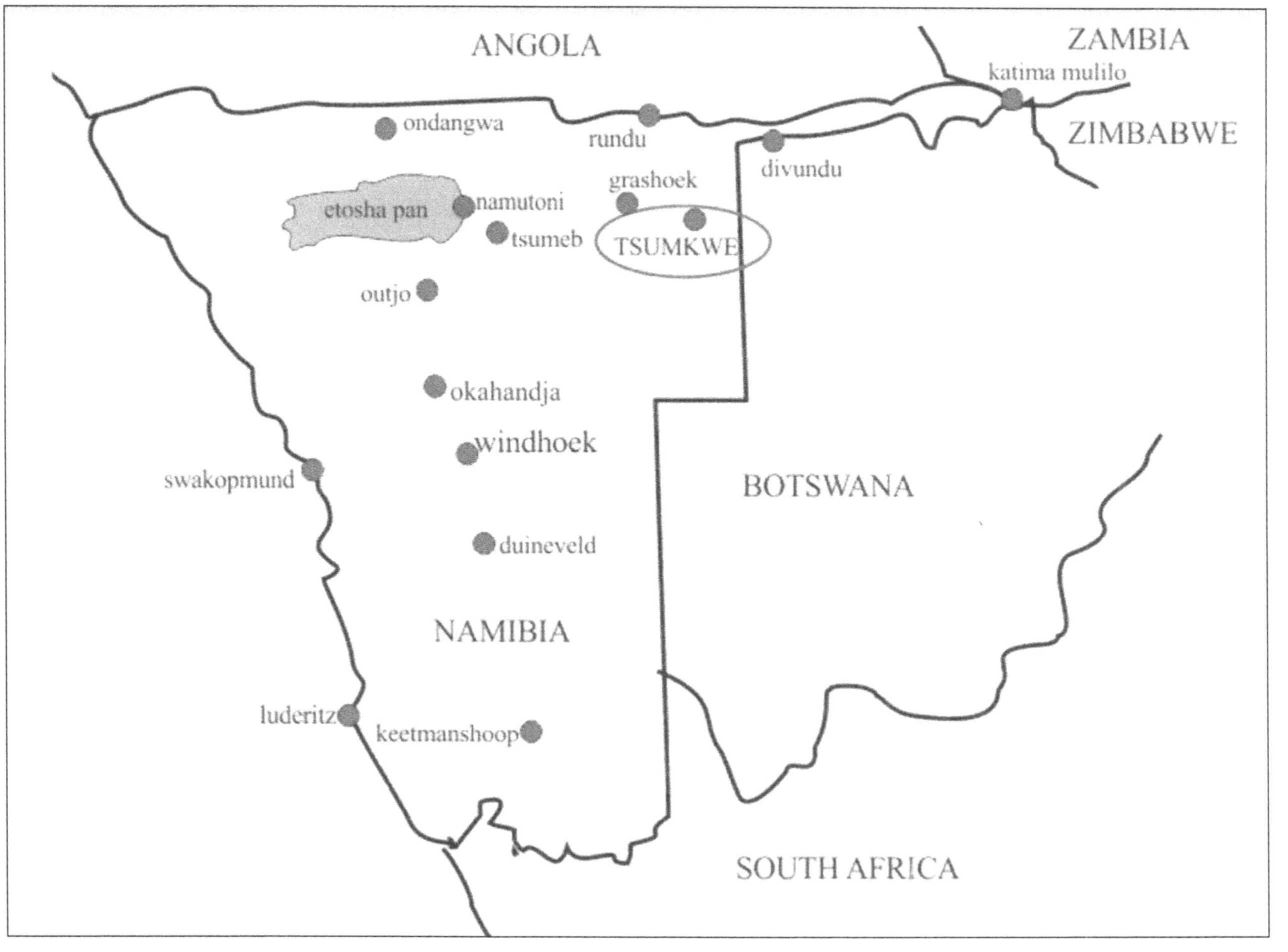

Figure 54. The location of Tsumkwe settlement in Namibia (after Sarantou 2018).

role in making fire (Figure 55). The thin branches are half-perforated along the horizontal axis and to initiate a fire, a second stick is vigorously brushed against each hole. Grass is usually put around the friction area to quicken the fire-starting process (Omandumaba interviews). The preparation of *Sansevieria pearsonii* fleshy stems for making strings used for bird traps was also demonstrated (Figure 56).

Carefully formulated questions for ethnographic purposes might be subjective. In this study, for example, the informants gave responses only about plants that are found in their immediate proximities but did not mention some, though it became apparent they use them occasionally. Plants found in the charcoal assemblage at Leopard Cave and around the site such as *Vangueria infausta* could be accounted for only when specific questions were asked.

Finally, the San people (who were initially nonamdic) today come to Omandumba from Tsumkwe, for short, alternated periods of time, source almost all their vegetal materials locally as do the Damara of Tubusis.

The results from the comparison of the wood used between the two ethnic groups living in the same environment are significantly interesting (Table 17); The members of the Damara community are more eclectic when it comes to using wood as fuel. They collect almost all trees and shrubs available in their environment for making fire. Even though they do not fell trees for firewood, it is apparent that they do not harvest from what they deem trees with edible fruit or with fruit used in a variety of activities. The Ju/'Hoansi on the other hand are very selective and tend to omit the use of certain plants for fuel. All *Boscia* species, for example, are avoided as they are associated with taboos. Some informants highlighted that using such plants in the fire may trigger severe respiratory illness to those inhaling the smoke. Some persons of the Damara community in Tubusis simply mention that they prefer not to use certain wood types in the fire by tradition, e.g. they were considered as bad or forbidden by earlier generations as they were associated with bad luck and misfortune.

Table 16. Use of woody plants by the San community in Omandumba

Family	Scientific name	Common name	Uses	Presence in the archaeobotanical record
Burseraceae	*Commiphora wildii* Merxm.	Namibian Myrrh	• Wood: bark-snake repellent in houses, associated insects used for arrow poison	absent
Capparaceae	*Boscia albitrunca* (Burch.) Gilg & Gilg-Ben.	shepherd tree	• Leaves: are crushed and used as anti-inflammatory medication for the skin. *Roots: sliced and boiled/ steamed against chest infections . • The wood is toxic and not to be used for fuel. The smoke from the wood is capable of wiping out a whole family-informant. *Fruit: eaten fresh	charcoal and seeds
Celastraceae	*Gymnosporia senegalensis* (Lam.) Loes	confetti tree	• Branches: for making arrows in association with bluegrass (*Bothriochloa bladhii)* which is used to fix the arrow.	absent
Combretaceae	*Combretum apiculatum Sond. subsp. Apiculatum*	red bushwillow	• Young branches are used to build the hut "*!xhabusi*". • Seeds: used in body ornamentation, usually in combination with seeds from other plants.	charcoal and seeds
Euphorbiaceae	*Croton gratissimus* Burch.	lavender croton	• Wood: firewood and building material.*Roots -boiled to treat chest, back and joint pains.	seeds
Fabacaeae-Caesalpinoideae	*Adenolobus garipensis* (E.Mey.) Torre & Hillc	butterfly leaf	• Twigs: cut and used for making necklaces	absent
Fabacaeae-Caesalpinoideae	*Mundulea sericea* (Willd.) A. Chev.	cork bush	• Young branches: used to build the hut "*!xhabusi*" and are usually joined to those of *Terminalia prunioides, Acacia erubescens* and Sensevieria spp.	possibly present (charcoal)
Fabacaeae-Mimosoideae	*Acacia erubescens* Welw. ex Oliv.	blue thorn acacia	• Young branches: used to build the hut "*!xhabusi*". • Seeds are used in body ornamentats.	possibly present
Fabacaeae-Mimosoideae	*Acacia mellifera* (Vahl) Benth.	black thorn acacia	• Wood: firewood	possibly present
Fabacaeae-Mimosoideae	*Acacia tortilis* (Forssk.) Hayne	umbrella thorn acacia	• Wood: firewood. *Root-hollowed out and covered with animal skin to manufacture quivers.	possibly present
Fabacaeae-Mimosoideae	*Dichrostachys cinerea* (L.) Wight & Arn.	sicklebush	• Stem: carved into walking and digging sticks. • Gum: as bird trap or bait. *Seeds: mixed with other plant materials and used to make body ornaments.	present
Olacacea	*Ximenia americana* L.	yellow plum	• Fruit: eaten fresh. *Kernels: used as body ornamentation, burnt for inhalation or crushed and applied to the nostrils to cure headaches.	charcoal and seeds
Rubiaceae	*Vangueria infausta* Burch.	wild medlar	• Roots: crushed and taken orally for heart ailments	seeds
Tiliaceae	*Grewia flava* DC.	velvet raisin	• Tree provides shade. *Wood: firewood. *Branches: used for bows. *Leaves: used for making bows.*Leaves: dried and taken as an infusion or soaked in water to wde off evil spirits. *Fruit: eaten raw.	charcoal and seeds

Figure 55. Fire stick made from the Mongongo tree (*Schinziophyton rautanenii*). Photo credit: M. Mvimi.

Figure 56. A San man working *Sansevieria pearsonii* leaves into a string for bird trapping. Photo credit: M. Mvimi.

Table 17. Comparison of the wood species used by the Damara and San communities

Family	Species	San	Damara
Burseraceae	*Commiphora wildii*	X	X
Capparaceae	*Boscia albitrunca*		X
Capparaceae	*Boscia foetida*		X
Celastraceae	*Gymnosporia senegalensis*		X
Combretaceae	*Combretum apiculatum*		
Combretaceae	*Terminalia prunioides*		X
Euphorbiaceae	*Croton gratissimus*	X	
Euphorbiaceae	*Spirostachys africana*		X
Fabaceae-Caesalpinoideae	*Adenolobus garipensis*	X	X
Fabaceae-Caesalpinoideae	*Mundulea sericea*	X	X
Fabaceae-Mimosoideae	*Vachelia erioloba*		X
Fabaceae-Mimosoideae	*Vachellia erubescens*	X	X
Fabaceae-Mimosoideae	*Vachellia karoo*	X	X
Fabaceae-Mimosoideae	*Senegalia mellifera*	X	X
Fabaceae-Mimosoideae	*Vachelia tortilis*	X	X
Fabaceae-Mimosoideae	*Dichrostachys cinerea*		X
Fabaceae-Mimosoideae	*Faidherbia albida*	X	X
Olacaceae	*Ximenia americana*	X	X
Rhamnaceae	*Ziziphus mucronata*		X
Rubiaceae	*Vangueria infausta*	X	X
Sterculiaceae	*Sterculia africana*	X	X
Tiliaceae	*Grewia flava*		X

5

Results

5.1. Description of identified taxa

The first and major step in this study is the identification of the botanical remains collected at the three sites of Leopard Cave, Geduld and Toteng. By the application of anatomical criteria and the use of reference collections held in several laboratories as well as specifically collected in the framework of this thesis, 20 different taxa of ligneous species from wood charcoal could be botanically determined (Table 18). While it has been possible in some cases to determine the species, other fragments could only be determined at the family stage.

Below follows a presentation of the identified taxa and the anatomical criteria used for these identifications. The descriptions are illustrated by anatomical photographs obtained through the use of a Scanning Electronic Microscope (SEM). The description of wood anatomical criteria is followed by two figures illustrating a selection of desiccated and charred seed/fruits remains from Leopard Cave and Geduld (Figure 57, Figure 58). These remains were identified by the study of their general morphology compared, as was the case with the wood, to specimens in reference collections and descriptions in specialised atlases.

The presentation of the taxa is followed by the exposition of the quantitative results of the study.

Table 18. Comparison of the anatomic characteristics of the archaeological charcoal

Family	Charcoal family and type	Vessel arrangement and pitting						Ray width and height				
		solitary	radial multiples and/or clusters	Mixed(solitary, in pairs or multiples)	diameter of vessel lumina(µm)	scalariform	alternate	1–2 seriate	1–3 seriate	1–5 seriate	2–5/6 seriate	Ray height (µm)
Fabaceae/ Mimosoideae	*Acacia/Dichrostachys*			x	61		x		x			250
	Acacia/*Dichrostachys* 2	x			125		x				x	450
	Acacia type 1			x	200		x				x	460
	Acacia type 2	x			50–100		x		x			560
	Acacia type 2			x	49		x			x		360
	Acacia type 4	x			60		x			x		240
	Acacia type 5		x		50–100	x				x		370
	Acacia type 6				80		x			x	x	200
	Acacia type 7	x			100		x				x	430
	Acacia type 7	x			100		x			x		340
	Fabaceae	x			50–100		x	x				120
Fabaceae/ Caesalpinoideae	*Colophospermum mopane*		x	x	50–100		x		x			350
	Peltophorum africanum	x			50–100		x		x			150
Capparaceae	*Boscia/Maerua*	dentritic			<=50		x			x		300
	Boscia/Maerua		x		50–100		x		x			690
	Boscia type	nested			<=50		x		x			200
	Capparaceae	x			<=50		x		x			430
Combretaceae	*Terminalia prunoides* type		x		50–100		x	x				200
	Combretum molle	x			50–100		x					1,1mm
	Terminalia type				50–100		x	x				330

*C. mopane has similarities with Vachellia/Senegalia type 2; 1*The crystals are in Idioblasts and also present is the gelatinous fibre in vessels; 2* Crystals in enlarged cells; 3* Gum and other deposits in ray and vessel cells; 4*Diffuse parenchyma and the presence of enlarged cells.
Boscia and Maerua differ in that those from Leopard Cave present no crystals and no tension wood but those from Geduld do.
Boscia type at Geduld, on the other hand, laacks crystals but has resin canals

Ray compostion							Parenchyma						Mineral inclusions						Other
procumbent	upright	Heterocellular (mixed)	perforated ray cells	parenchyma associated with the rays	Radial canals	septate fibres	vasicentric	banded	aliform	confluent	apotracheal	absent or rare	crystals in parenchyma cells	crystals in ray cells	crystals in fibre cells	Silica in ray cells	Druses	inclusions additional information	Included phloem
x						x	x	x		x			x				x*		
x			x					x		x				x			x		
x							x			x			x					1*	
x							x						x		x				
x															x				
x							x	x					x		x				
x						x	x	x					x		x				
		x											x	x			x		
x							x			x				x					
		x				x	x		x	x				x					
x							x		x					x			x	2*	
x							x		x*				x						
x				x			x	x	x				x						
x				x							x			x					
	x			x							x								
x					x		x											3*	
		x										x						4*	
x											x		x				x		
		x					x			x				x		x	x		x
x																			

Geduld: FABACEAE - *Vachellia/Senegalia /Dichrostachys type*

IAWA Feature Code: 2,5,9,10, 13,22,24,41,47,53,56,61,65,79,83,97,102,104,115,136,142

- Growth ring boundaries absent
- Wood diffuse porous
- Vessels mostly solitary (more than 90 percent) and rarely in pairs or radial multiples
- Solitary vessel outline rounded
- Simple perforation plates
- Alternate and quasi-scalariform intervessel pitting
- Minute intervessel pits: <= 4µm
- Mean tangential diameter of vessel lumina: 50–100µm
- 5–20 vessels per square millimetre
- Mean vessel element length: 350–800µm
- Tyloses present in vessels(a)
- Fibres with simple to minutely bordered pits
- Septate fibres present (f)
- Very thick-walled fibres
- Paratracheal axial parenchyma present
- Axial parenchyma vasicentric and confluent
- 2–4 cells per parenchyma strand
- Rays 1–3 cells wide
- Ray height: >1mm
- All ray cells procumbent
- 4–12/ rays per millimetre
- Prismatic crystals in fibre cells
- Prismatic crystals in chambered parenchyma cells(c)
- More than one crystal of about the same size per cell or chamber
- Tension wood present

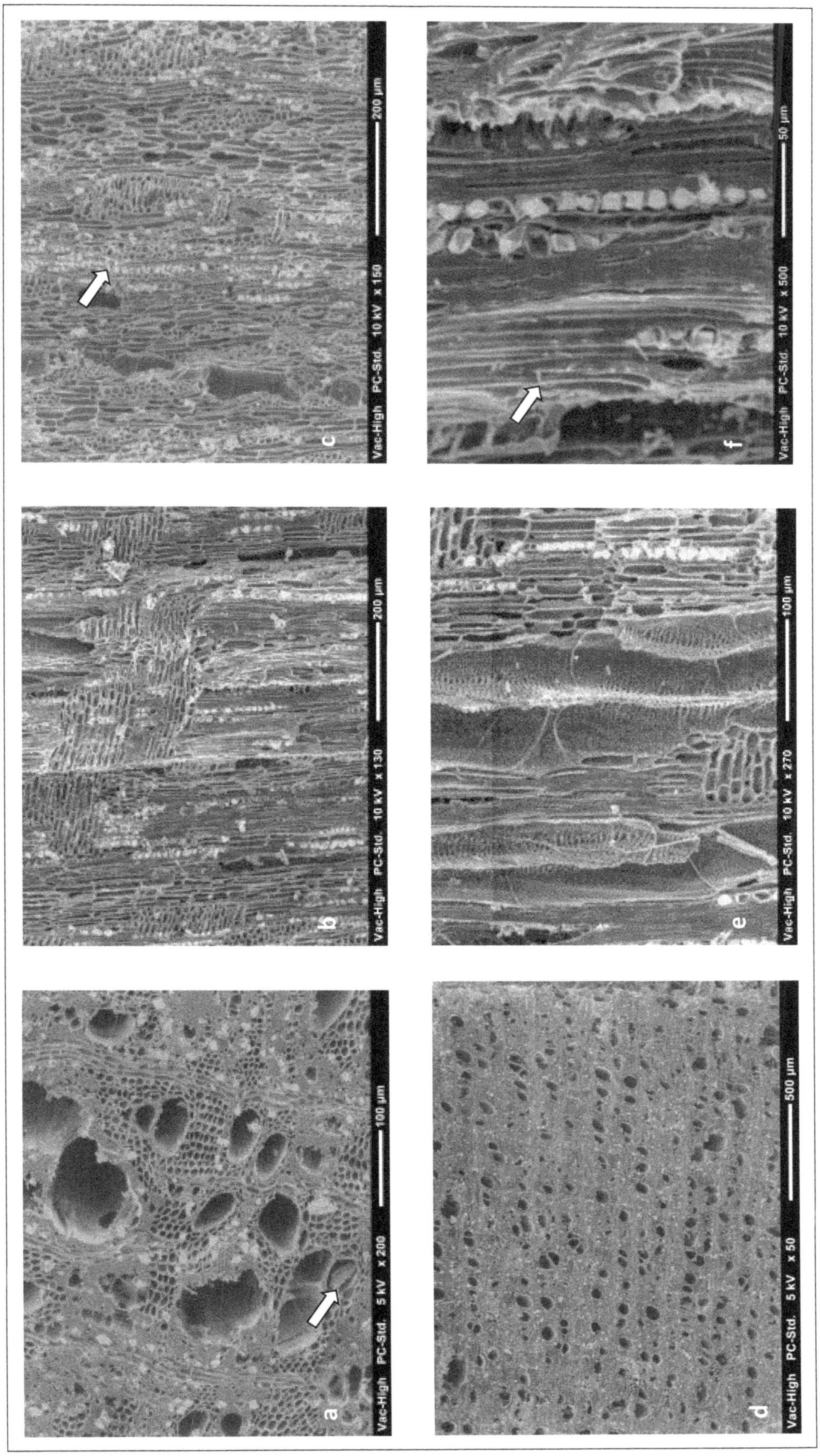
a
Vac-High PC-Std. 5 kV x 200
100 µm
b
Vac-High PC-Std. 10 kV x 130
200 µm
c
Vac-High PC-Std. 10 kV x 150
200 µm
d
Vac-High PC-Std. 5 kV x 50
500 µm
e
Vac-High PC-Std. 10 kV x 270
100 µm
f
Vac-High PC-Std. 10 kV x 500
50 µm

Geduld: FABACEAE - *Vachellia/Senegalia* type 1

IAWA Feature Code: 2,5,9,10,13,22,24,42,48,60,61,79,83,98,102,104,110,115, 136,141,156,181

- Growth ring boundaries absent
- Wood diffuse porous
- Vessels mostly solitary (more than 90%) and rarely in pairs or radial multiples
- Solitary vessel outline rounded
- Simple perforation plates
- Alternate intervessel pitting
- Minute intervessel pits: <= 4µm
- Mean tangential diameter of vessel lumina: >200µm
- 20–40 vessels per square millimetre
- Gums and other deposits in heartwood vessels (c)
- Vascular tracheids present
- Helical thickenings in ground tissue fibres
- Fibres with simple to minutely bordered pits
- Thin -to thick-walled fibres
- Paratracheal axial parenchyma present
- Axial parenchyma vasicentric and confluent
- 5–8 cells per parenchyma strand (b)
- Rays 1–5 cells wide
- Ray height: >1mm
- All ray cells procumbent
- Sheath cells present (c)
- 4–12/ rays per millimetre
- Prismatic crystals in ray-fibre and parenchyma cells (a and c)
- Prismatic crystals in non-chambered axial parenchyma cells
- Druses in ray parenchyma cells
- Crystals in enlarged cells
- Tension wood present (e)

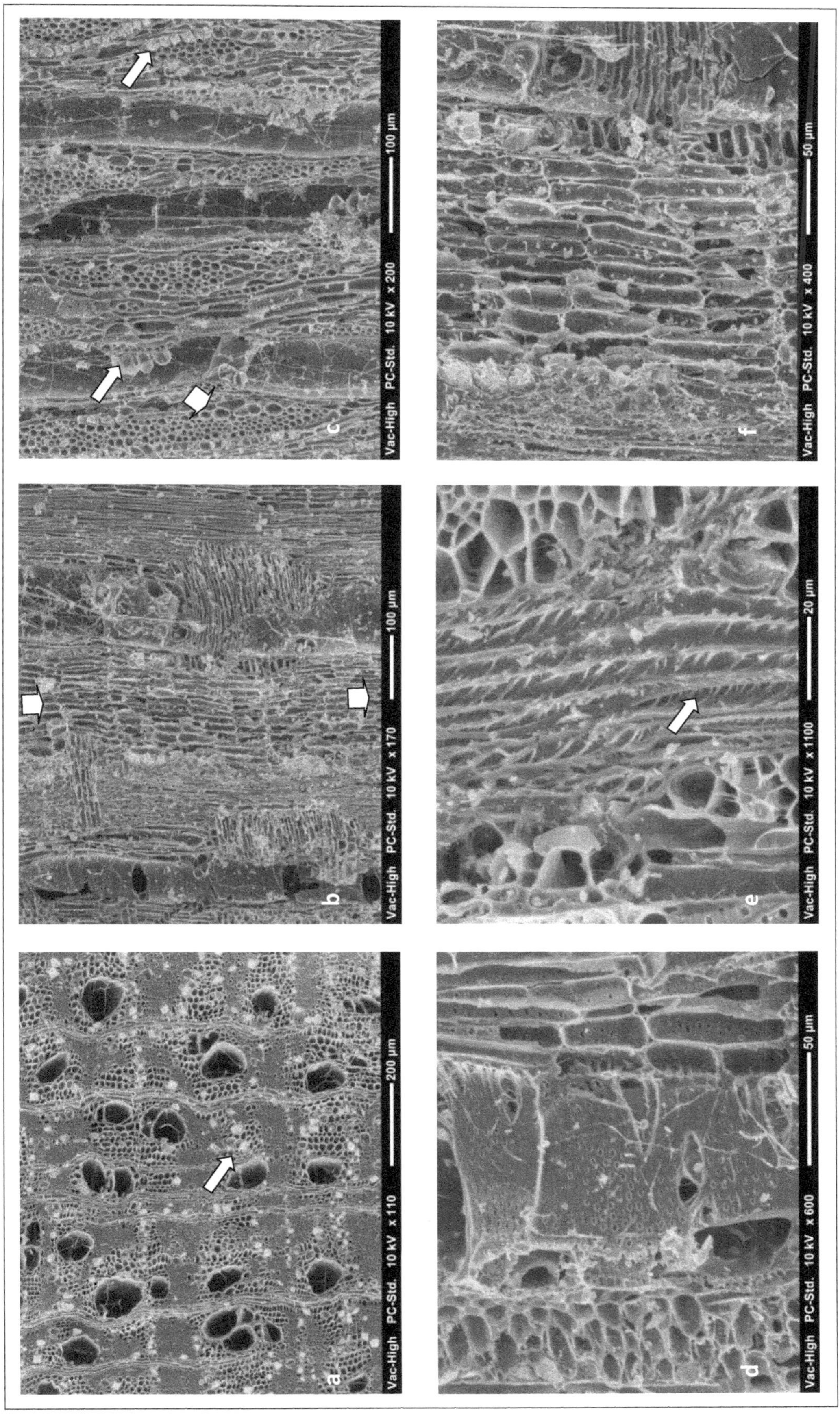

a : transversal; b,d and f: radial; c and e: tangential

Geduld: FABACEAE type

IAWA Feature Code: 2,5,9,10,13,22,24,41,46,56,64,65,69,79,83,94,97,102,104, 115,136,143,144,156,181

- Growth ring boundaries absent
- Wood diffuse-porous
- Vessels mostly solitary (more than 90%) and rarely in pairs or radial multiples
- Solitary vessel outline rounded
- Simple perforation plates
- Alternate intervessel pitting
- Minute intervessel pits: <= 4µm
- Mean tangential diameter of vessel lumina: 50-100µm
- =<5 vessels per square millimetre
- Tyloses are common in vessel elements (b and d)
- Helical thickenings in ground tissue fibres
- Septate fibres present (c)
- Thin -to thick-walled fibre walls
- Paratracheal axial parenchyma present(d)
- Axial parenchyma vasicentric and confluent (a and d)
- Over 8 cells per parenchyma strand
- Rays 1–3 cells wide
- Ray height: >1mm
- All ray cells procumbent
- 4–12/ rays per millimetre
- Prismatic crystals in ray-fibre cells
- Druses in ray parenchyma cells (f)
- Helical thickenings in ground tissue fibres (f)

a and d: transversal, b and e: radial, c and f: tangential

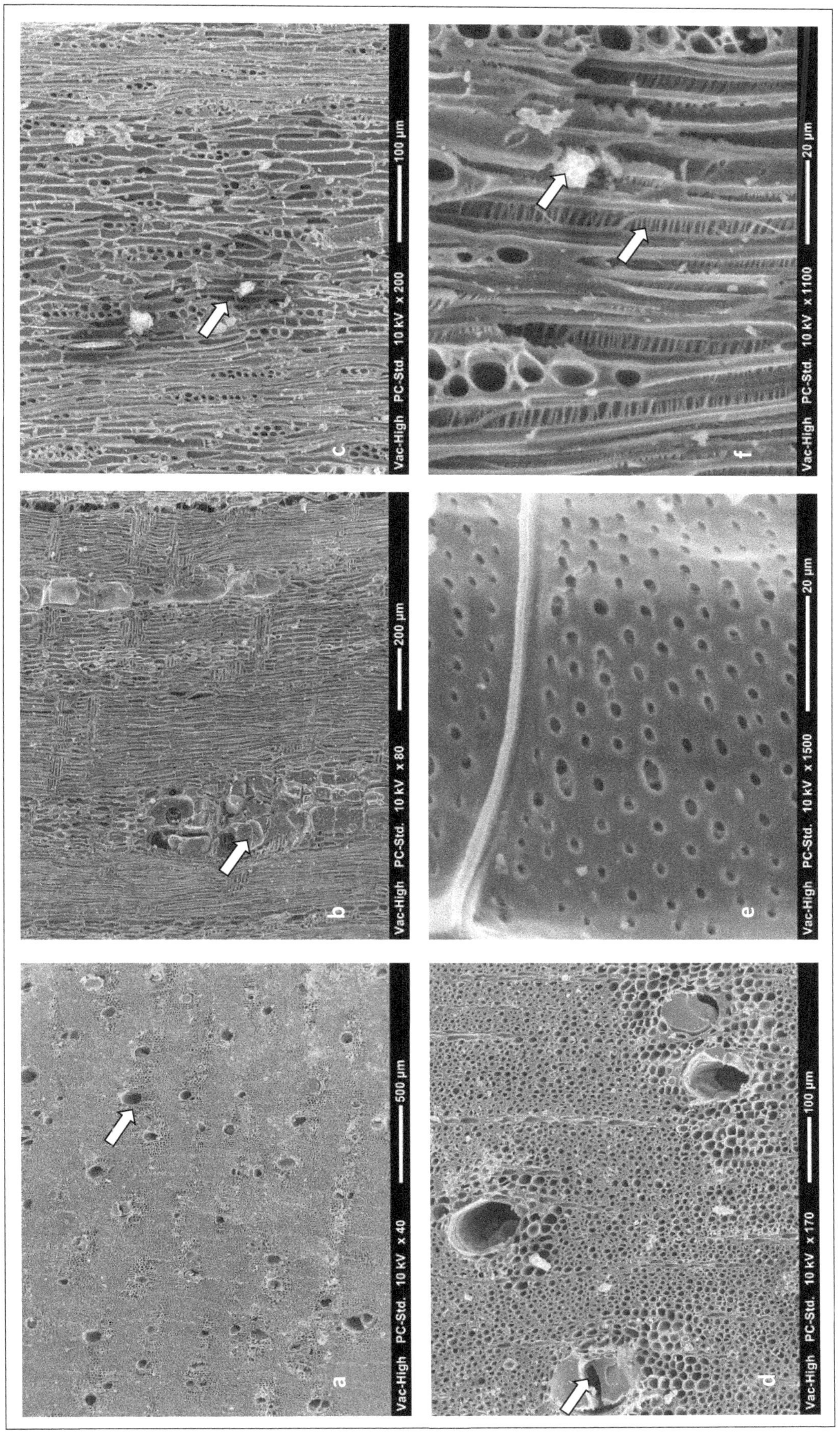

Geduld: FABACEAE - *Vachellia/Senegalia* type 2

IAWA Feature Code: 2,5,9,10,13,22,24,41,47,56,84,98,10,107,115,136,142,181

- Growth ring boundaries absent
- Wood diffuse porous
- Vessels mostly solitary (more than 90%) and rarely in pairs or radial multiples
- Solitary vessel outline rounded
- Simple perforation plates
- Alternate intervessel pitting
- Minute intervessel pits: <= 4µm
- Vestured pits present (f)
- Mean tangential diameter of vessel lumina: 50-100µm
- 5–20 vessels per square millimetre
- Tyloses and gum deposits in vessels
- Helical thickenings in ground tissue fibres
- Fibres very thick-walled
- Paratracheal axial parenchyma present
- Axial parenchyma unilateral paratracheal is some areas
- Over 8 cells per parenchyma strand
- Rays 1–5 cells wide
- Ray height: >1mm
- Body ray cells procumbent with mostly 2–4 rows of upright and/or square marginal cells
- Minute alternate inter-ray pitting: <= 4µm
- 4–12/ rays per millimetre
- Minute alternate pitting in parenchyma cells: <= 4µm
- Prismatic crystals in fibre cells (especially ray-fibre cells), procumbent rays
- Prismatic crystals in chambered axial parenchyma cells
- Tension wood present
- Idioblasts present
- Gelatinous fibre in vessels

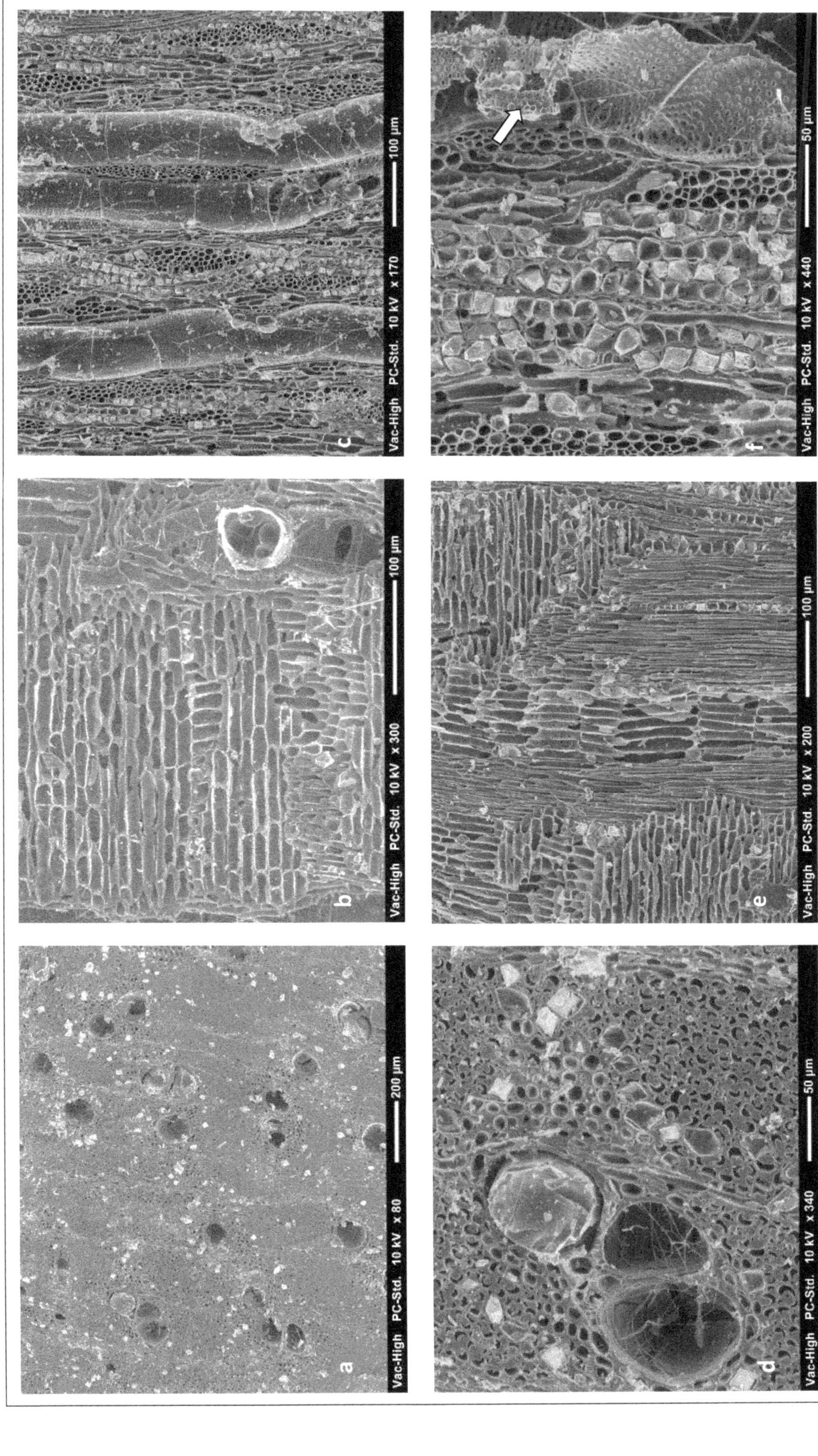

a and d: transversal, b and e: radial, c and f: tangential

Leopard Cave: FABACEAE - *Vachellia/Senegalia* type 2

IAWA Feature Code: 2,5,9,13,22,24,42,47,75,79,83,96,98,102,104,115,136,156,181

- Also, Vachellia/Senegalia type 2

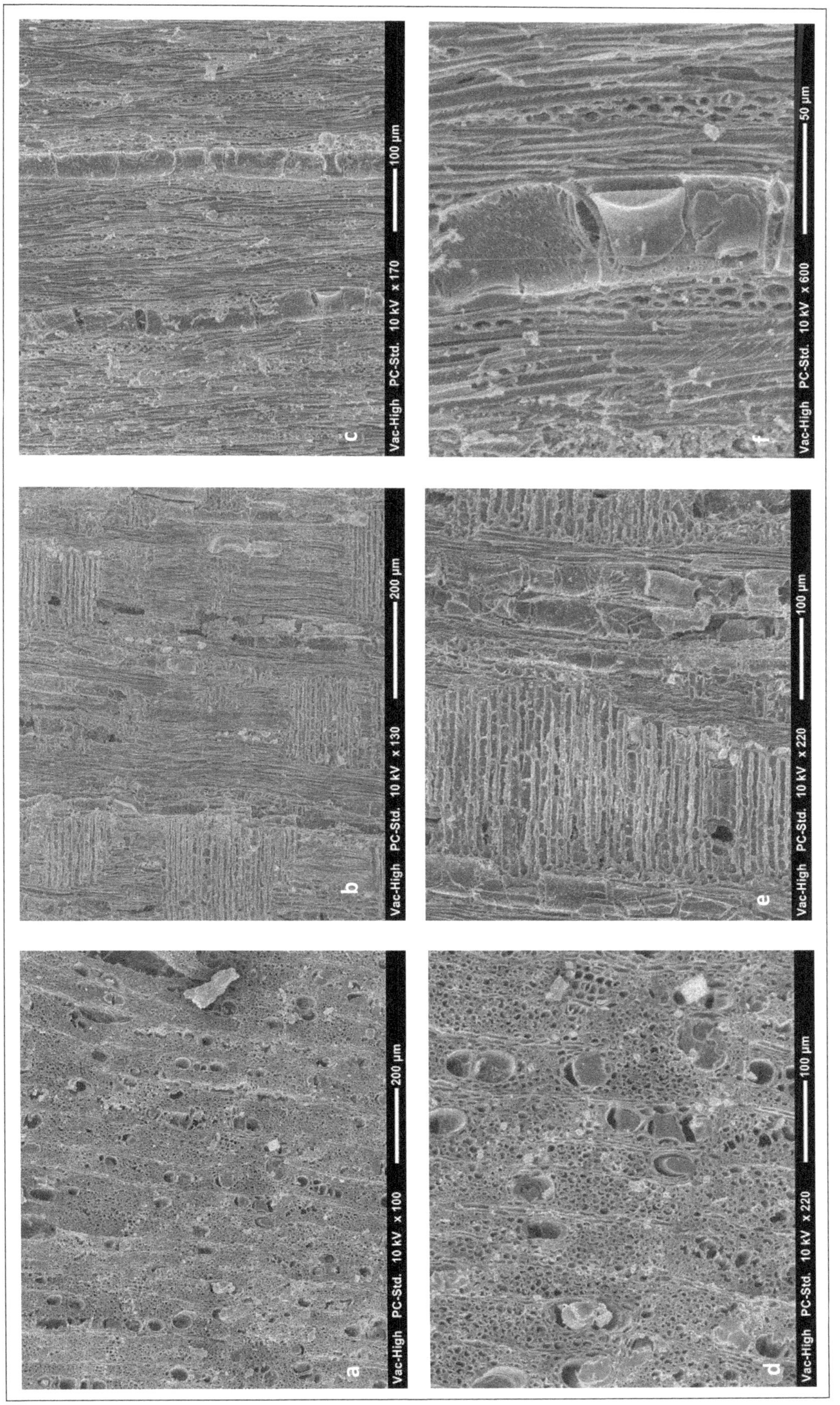

a and d: transversal, b and e: radial, c and f: tangential

Geduld: FABACEAE - *Vachellia/Senegalia* type 4

IAWA Feature Code: 2,5,9,10,13,22,24,41,48,56,84,85,98,102,104,115,136,142,181

- Growth ring boundaries absent
- Wood diffuse porous
- Vessels mostly solitary (more than 90%) and rarely in pairs or radial multiples
- Solitary vessel outline rounded
- Simple perforation plates
- Alternate intervessel pitting
- Small intervessel pits: 4–7µm
- 20–40 vessels per square millimetre
- Non-septate fibres present
- Thin -to thick-walled fibre walls
- Paratracheal axial parenchyma present
- Axial parenchyma bands more than 3 cells wide
- Over 8 cells per parenchyma strand
- Minute alternate pitting in parenchyma cells: <= 4µm
- Rays 1–6 cells wide
- Ray height: >1mm
- All ray cells procumbent
- 4–12/ rays per millimeter
- Minute alternate inter-ray pitting: <= 4µm
- Prismatic crystals in fibre cells
- Prismatic crystals in chambered parenchyma cells
- Gum and other mineral inclusions present (d)

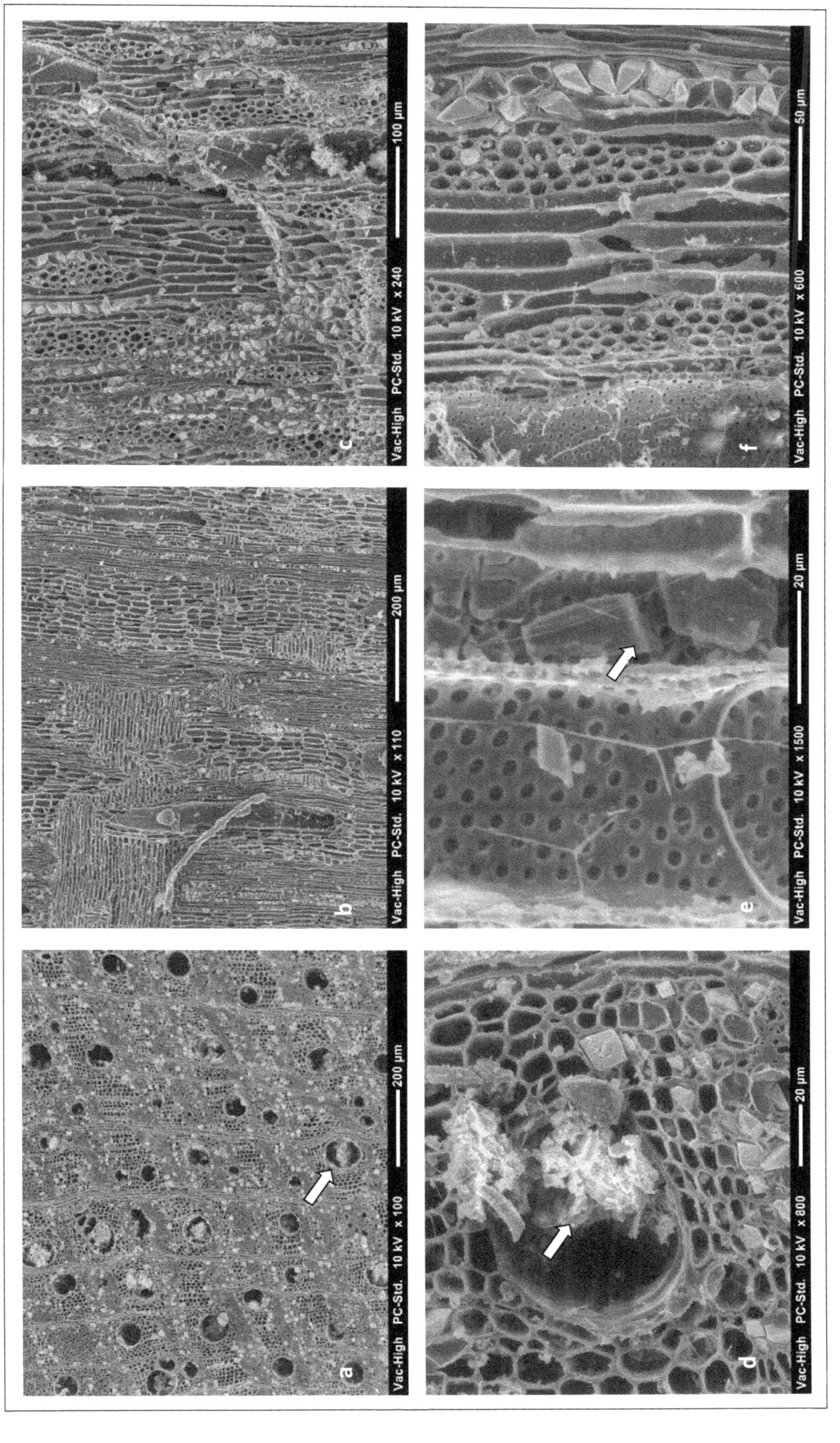

a and d: transversal, b and e: radial, c and f: tangential

Geduld: FABACEAE – *Vachellia/Senegalia* type 5

IAWA Feature Code: 2,5,7,10,13,22,24,27,41,48,49,60,85,97,102,104,115,136,142,181

- Growth ring boundaries indistinct or absent
- Wood diffuse porous
- Vessel groups in radial pattern
- Vessels in radial multiples of 4 or more common
- Vessels of two distinctive sizes(a)
- Angular vessel outline present: (a)
- Simple perforation plates (a)
- Alternate intervessel pitting
- Small to large intervessel pits
- Mean tangential diameter of vessel lumina: 50–100µm
- 40–100 vessels per square millimetre
- Vasicentric tracheids present
- Fibre pits common in both radial and tangential walls
- Septate fibres present
- Fibres very thick-walled (d)
- Paratracheal axial parenchyma present
- Axial parenchyma bands more than 3 cells wide
- Over 8 cells per parenchyma strand
- Minute alternate pitting in parenchyma cells: <= 4µm
- Rays 1–4 cells wide
- Ray height: >1mm
- All ray cells procumbent
- 4–12/ rays per millimetre
- Prismatic crystals in fibre cells (e)
- Prismatic crystals in chambered parenchyma cells
- Other plant deposit in parenchyma cells (a)

a and d: transversal, b and e: radial, c and f: tangential

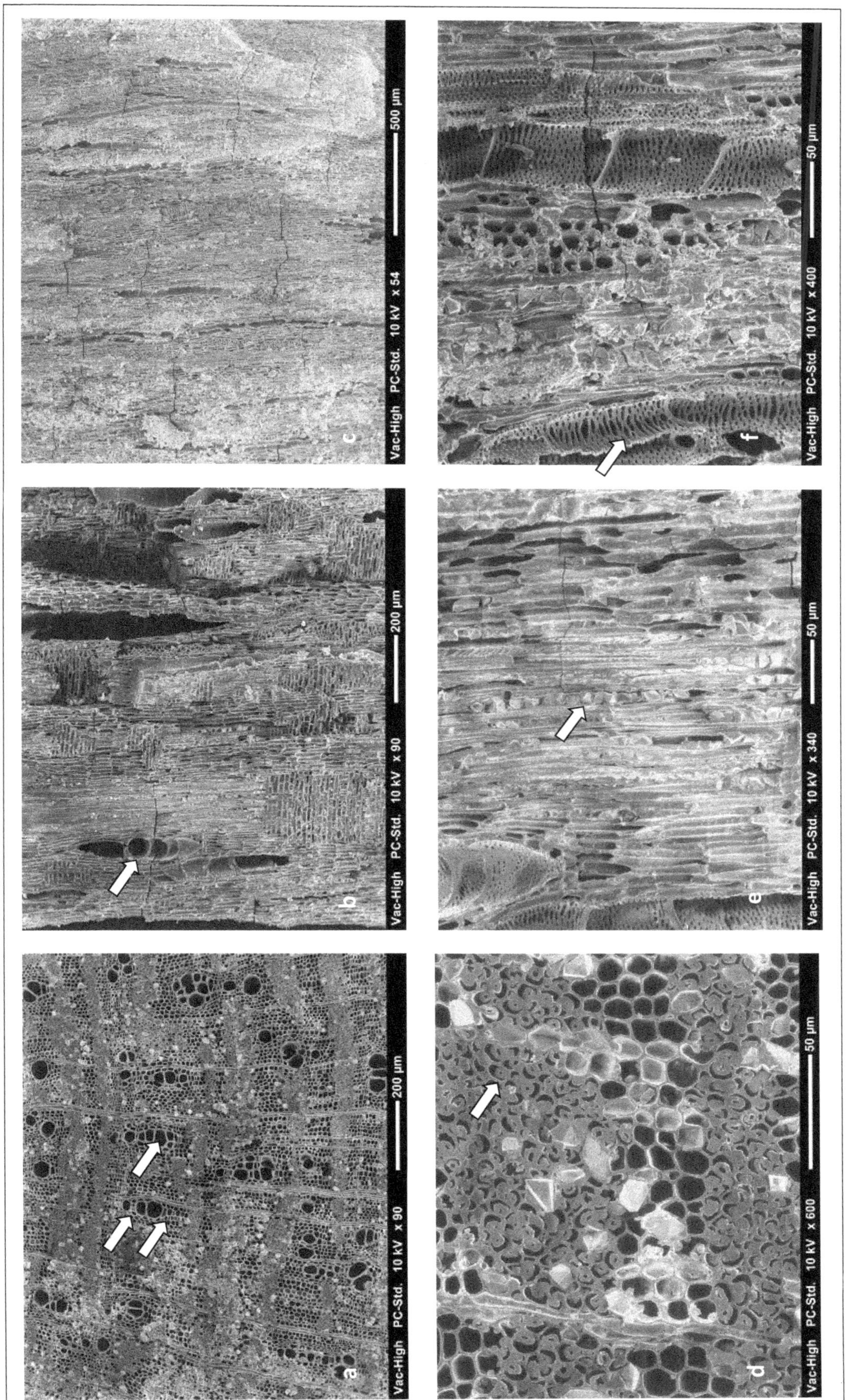

Geduld: FABACEAE - *Peltophorum africanum*

IAWA Feature Code: 2,5,9,10,13,22,24,27,41,47,60,65,83,97,102,106,115,136,141,181

- Growth ring boundaries absent
- Wood diffuse porous
- Vessels mostly solitary (more than 90%) and rarely in pairs or radial multiples
- Solitary vessel outline rounded
- Simple perforation plates
- Alternate intervessel pitting
- Minute intervessel pits: <= 4µm
- Mean tangential diameter of vessel lumina: 50–100µm
- 5–20 vessels per square millimetre
- Tension wood present (d)
- Septate fibres present
- Fibres thin walled
- Paratracheal axial parenchyma present
- Axial parenchyma confluent
- 5–8 cells per parenchyma strand
- Minute alternate pitting in parenchyma cells: <= 4µm
- Rays 1–2 cells wide
- Ray height: >1mm
- Body ray cells procumbent with one row upright and/or square marginal cells
- 4–12/ rays per millimetre
- Prismatic crystals in chambered parenchyma cells

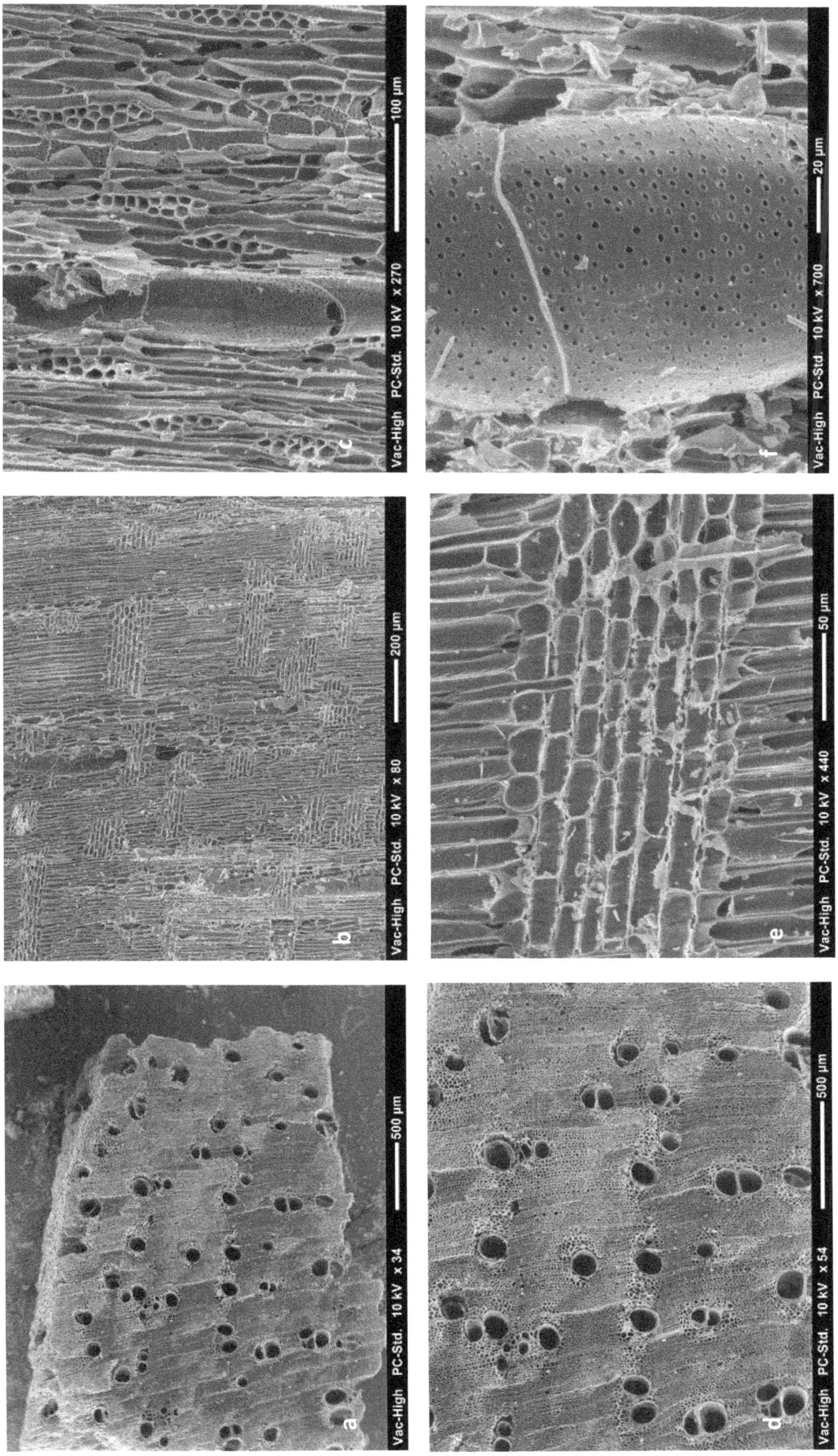

a and d: transversal, b and e: radial, c and f: tangential

Geduld: FABACEAE/CAESPALPINACEAE- *Colophospermum mopane*

IAWA Feature Code: 2,5,10,13,22,24,41,49,61,79,97,102,104,115,136,142p,156,181

- Growth ring boundaries absent
- Wood diffuse-porous
- Vessels in radial multiples of 4 or more common
- Solitary vessel outline rounded
- Simple perforation plates
- Alternate intervessel pitting
- Minute intervessel pits: <= 4µm
- Mean tangential diameter of vessel lumina: 50–100µm
- 40–100 vessels per square millimetre
- Fibres with simple to minutely bordered pits
- Fibres thin-walled
- Paratracheal axial parenchyma present
- Axial parenchyma vasicentric
- Over 8 cells per parenchyma strand
- Minute alternate pitting in parenchyma cells: <= 4µm
- Rays 1–3 cells wide
- Ray height: >1mm
- All ray cells procumbent
- 4–12/ rays per millimetre
- Prismatic crystals in fibre cells (especially ray-fibre cells), procumbent rays and
- Prismatic crystals in chambered axial parenchyma cells
- Crystals in enlarged cells (e right)
- Crystal sand present (e left)

a and d: transversal, b and e: radial, c and f: tangential

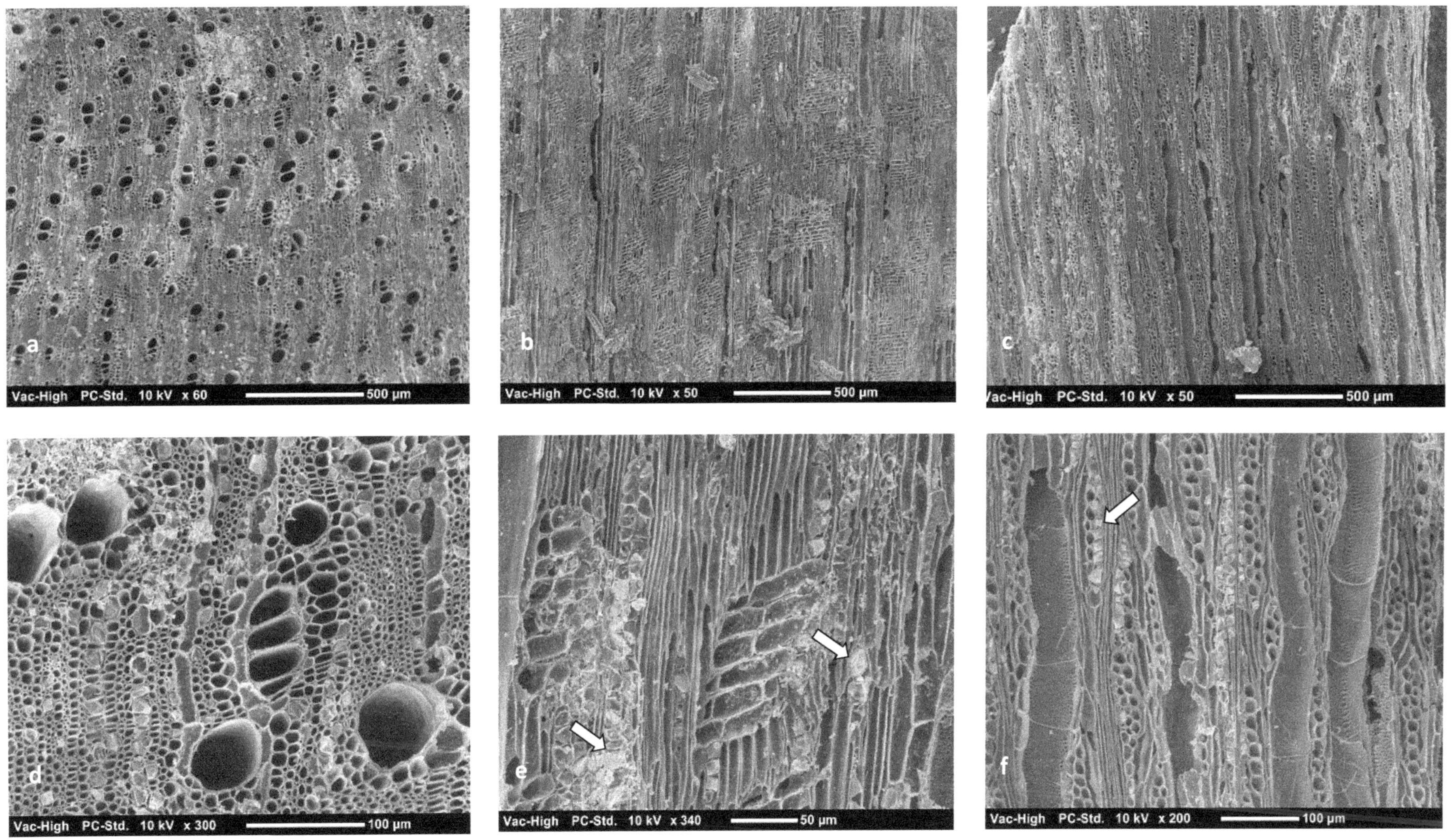

Toteng: FABACEAE - *Vachellia/Senegalia* type 7

IAWA Feature Code: 2,5,9,13,22,24,42,47,56,79,83,98,102,104,115,136,181

- Growth ring boundaries absent
- Wood diffuse porous
- Vessels exclusively solitary (90% or more)
- Solitary vessel outline rounded
- Simple perforation plates
- Alternate intervessel pitting
- Minute intervessel pits: <= 4μm
- Mean tangential diameter of vessel lumina: 100-200μm
- 5–20 vessels per square millimetre
- All ray cells procumbent
- 4–12/ rays per millimetre
- Tyloses and druses in vessels (present in vessel elements (c and b respectively)
- Thin-to thick-walled fibre walls
- Paratracheal axial parenchyma present
- Axial parenchyma vasicentric and confluent
- Over 8 cells per parenchyma strand
- Rays 1–5 cells wide
- Ray height: >1mm
- Prismatic crystals in ray-fibre cells
- Tension wood present

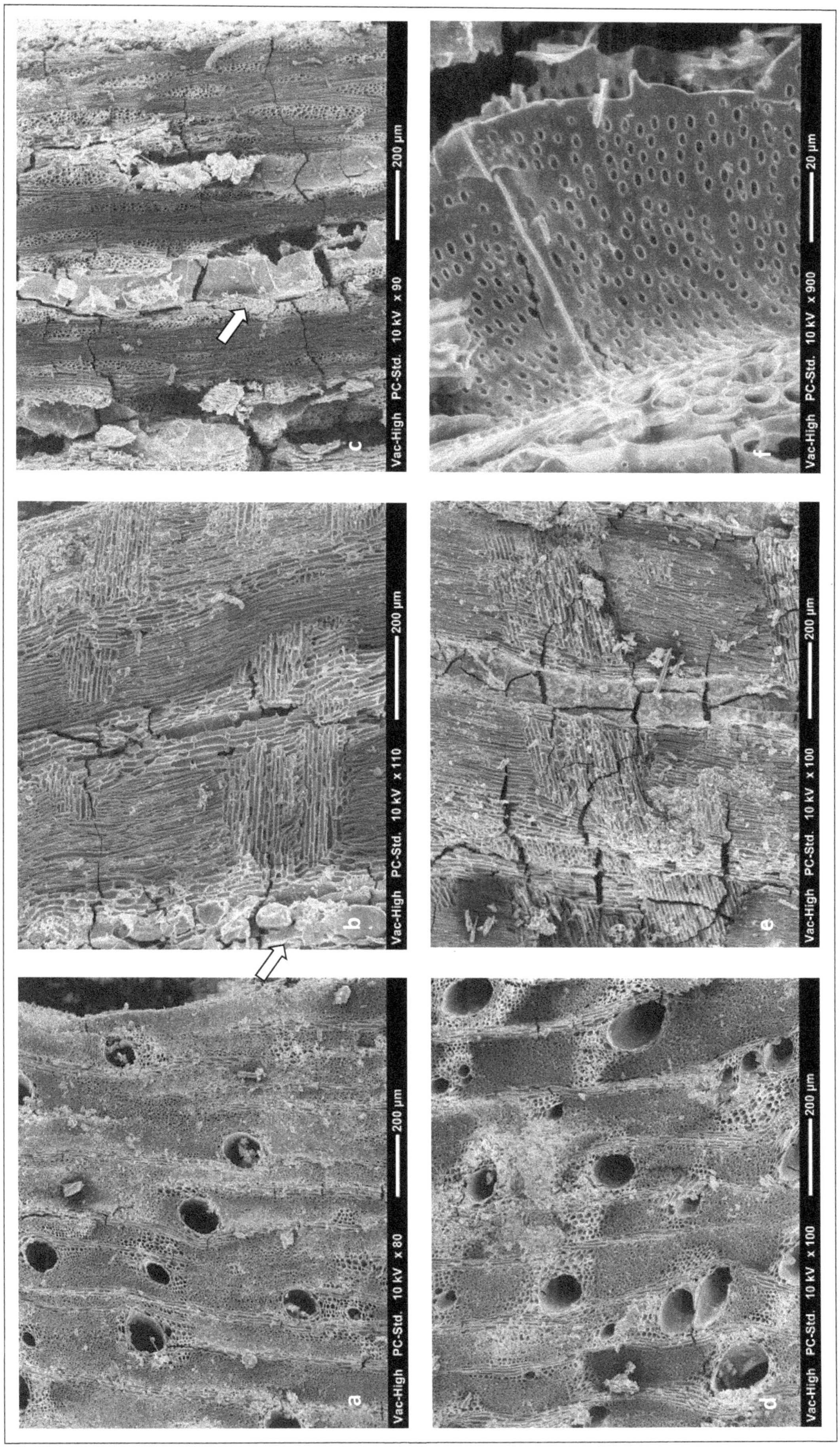

a and d: transversal, b and e: radial, c and f: tangential

Leopard Cave: FABACEAE - *Vachellia/Senegalia* type 7

IAWA Feature Code: 2,5,9,13,22,24,42,47,56,79,83,98,102,104,115,136,181

- Growth ring boundaries absent
- Wood diffuse porous
- Vessels exclusively solitary (90% or more)
- Solitary vessel outline rounded
- Simple perforation plates
- Alternate intervessel pitting
- Minute intervessel pits: <= 4µm
- Mean tangential diameter of vessel lumina: 100–200µm
- 5–20 vessels per square millimetre
- Tyloses present in vessel elements
- Septate fibres present (e)
- Thin-to thick-walled fibre walls
- Paratracheal axial parenchyma present
- Axial parenchyma vasicentric and confluent
- Over 8 cells per parenchyma strand
- Rays 1–5 cells wide
- Ray height: >1mm
- All ray cells procumbent
- 4–12/ rays per millimetre
- Prismatic crystals in ray-fibre cells (f)
- Druses in ray parenchyma cells (rounded and a crystal)-Toteng 1
- Druses in vessels
- Tension wood present in ground tissue fibres(f)

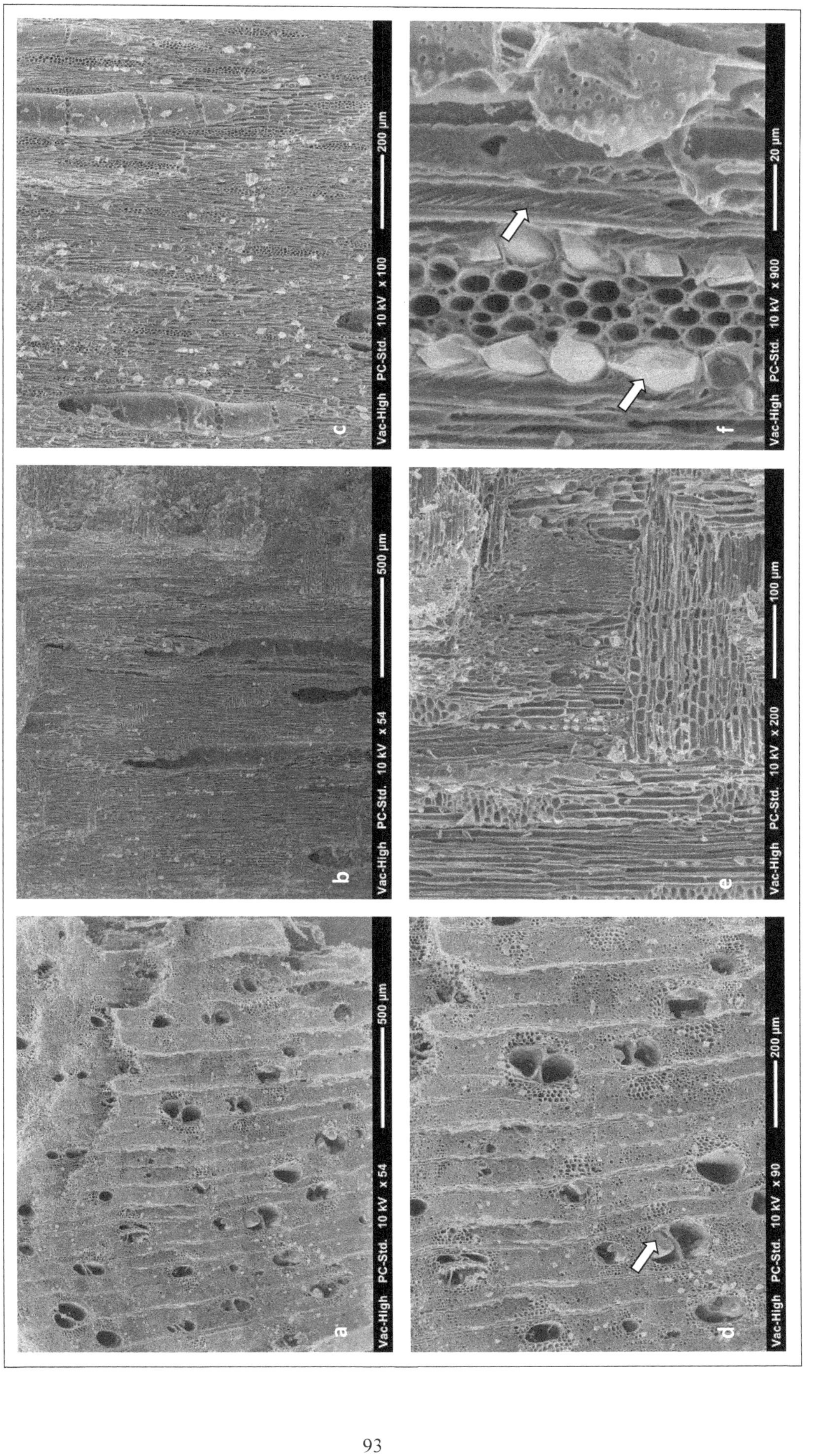

a and d: transversal, b and e: radial, c and f: tangential

Leopard Cave: FABACEAE - *Vachellia/Senegalia /Dichrostachys* type 2

IAWA Feature Code: 2,5,9,13,22,24,42,47,75,79,83,96,98,102,104,115,136,156,181

- Growth ring boundaries absent
- Wood diffuse porous
- Vessels exclusvely solitary (90% or more)
- Solitary vessel outline rounded
- Simple perforation plates
- Alternate intervessel pitting
- Minute intervessel pits: <= 4µm
- Mean tangential diameter of vessel lumina: 100–200µm
- 5–20 vessels per square millimetre
- Septate fibres present
- Tyloses present in vessel elements
- Thin-to thick-walled fibre walls
- Helical thickenings?
- Paratracheal axial parenchyma present
- Axial parenchyma vasicentric and confluent
- Over 8 cells per parenchyma strand
- Rays 1–5 cells wide
- Ray height: >1mm
- All ray cells procumbent
- 4–12/ rays per millimetre
- Minute alternate inter-ray pitting: <= 4µm (e)
- Prismatic crystals in fibre cells (especially ray-fibre cells) (f)
- Styloids/Elongated crystals(f)
- Druses in vessels (c)
- Tension wood present(f)

a: transversal; b and e: radial; c,d and f: tangential

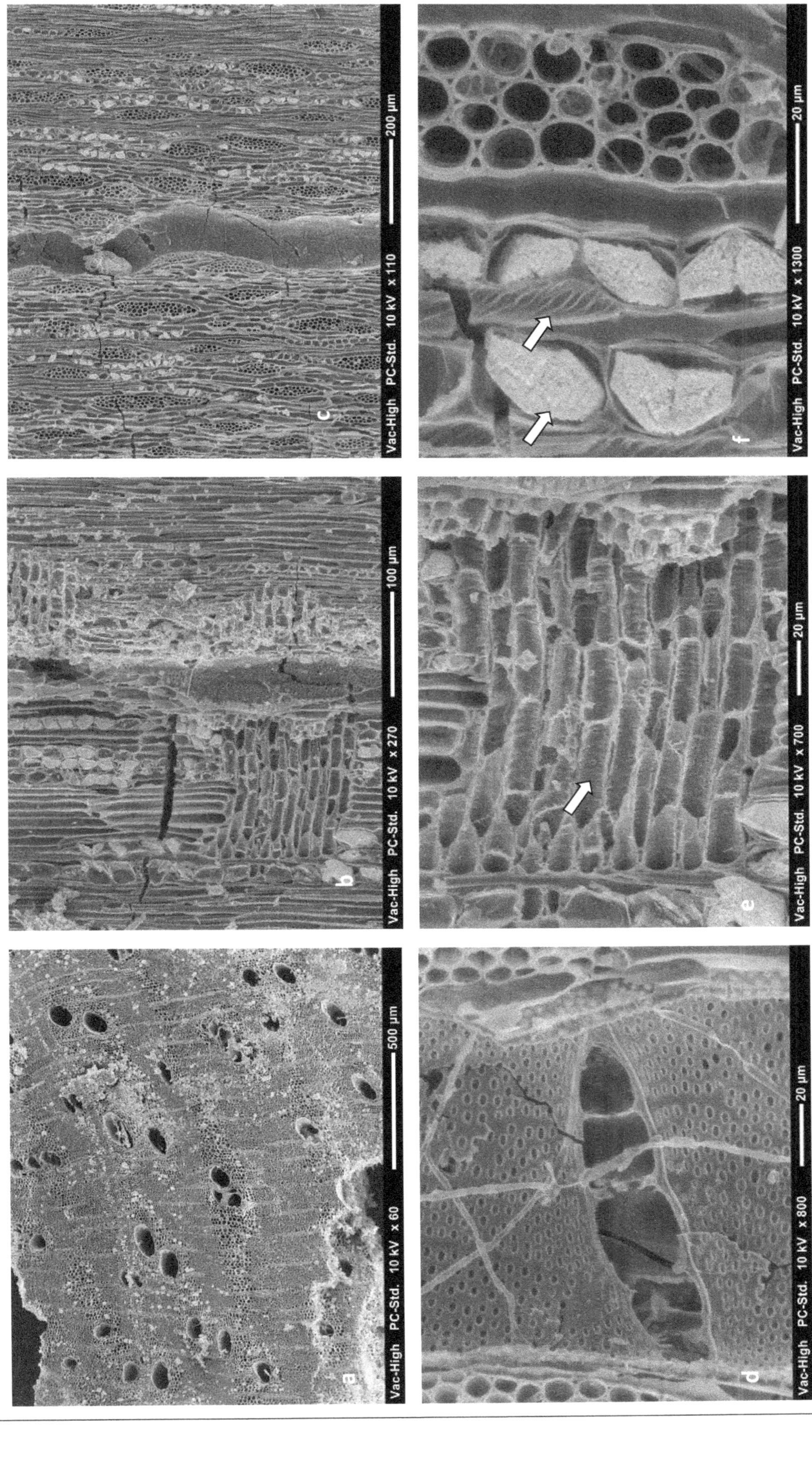

Geduld: CAPPARACEAE - *Maerua/Boscia*

IAWA Feature Code: 2,5,10,13,22,24,41,49,61,98,102,104,115,136,181

- Growth ring boundaries absent
- Wood diffuse porous
- Vessels in dendritic pattern
- Vessels in radial multiples of 4 or more common (up to 8)
- Solitary vessel outline rounded
- Vascular tracheids around vessel element (e)
- Simple perforation plates (e)
- Alternate intervessel pitting
- Minute intervessel pits: <= 4µm
- Mean tangential diameter of vessel lumina: 50–100µm
- 40–100 vessels per square millimetre
- Helical thickenings in ground tissue fibres
- Fibres also with simple to minutely bordered pits
- Fibres thin-walled
- Paratracheal axial parenchyma present
- Axial parenchyma also associated with ray cells
- Axial parenchyma in narrow bands
- Over 8 cells per parenchyma strand
- Minute alternate pitting in parenchyma cells: <= 4µm
- Rays 1–5 cells wide
- Ray height: >1mm
- Rays with procumbent, square and upright cells mixed throughout the ray
- 4–12/ rays per millimetre
- Prismatic crystals exclusively in ray cells (one crystal per cell) (f)
- Tension wood present (f)

a and d: transversal; b: radial; c,e and f: tangential

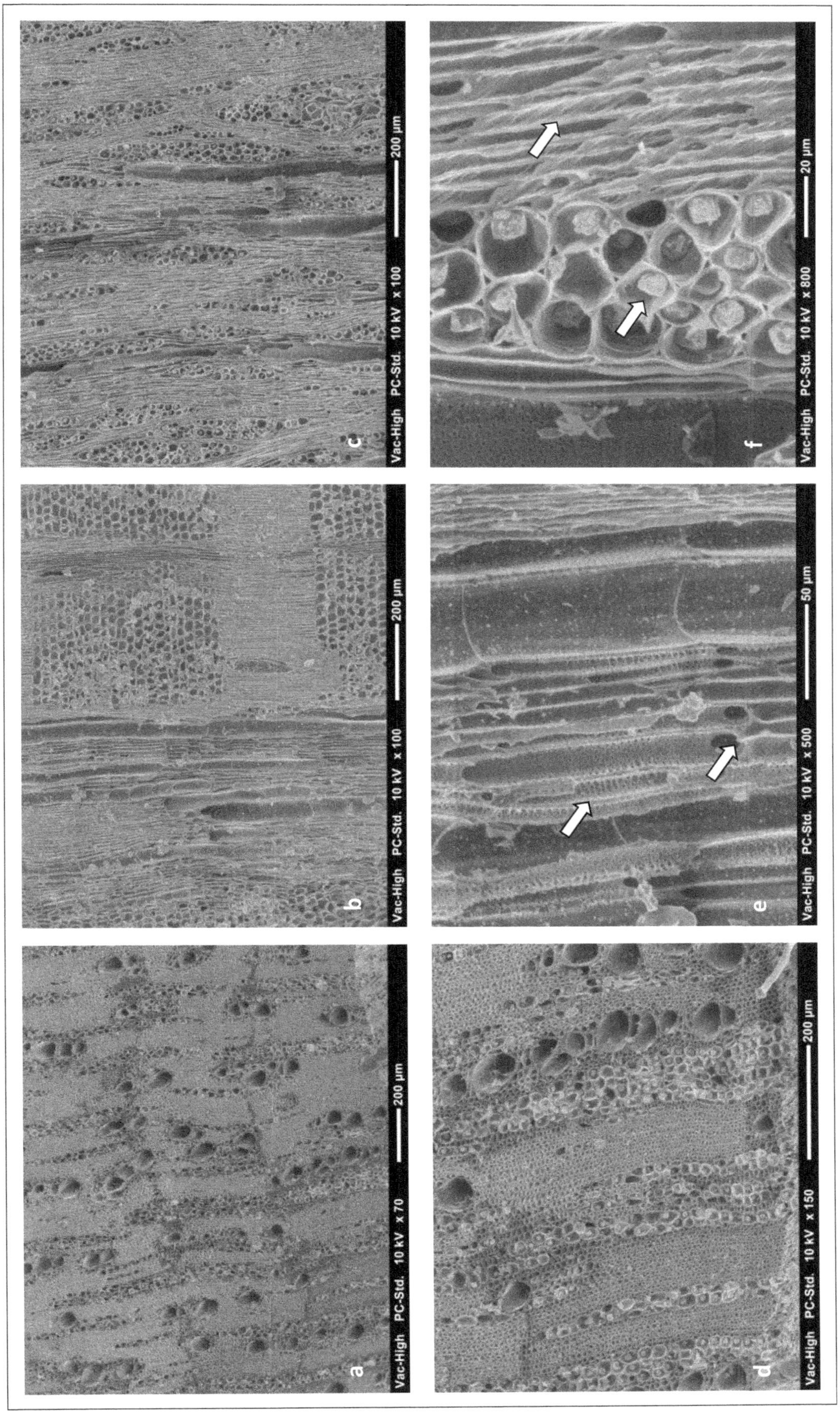

Leopard Cave: CAPPARACEAE - *Maerua/Boscia*

IAWA Feature Code: 2,5,8,10,13,22,24,41,49,61,79,83,86,97,102,104,115,136, 181 (comparisons with Geduld 1452 also probably maerua/boscia)

- Growth ring boundaries absent
- Wood diffuse porous
- Vessels in dendritic pattern
- Vessels in radial multiples of 4 or more common (up to 8)
- Solitary vessel outline rounded
- Simple perforation plates
- Alternate intervessel pitting
- Minute intervessel pits: <= 4µm
- Mean tangential diameter of vessel lumina: 50–100µm
- 40–100 vessels per square millimeter
- Fibres also with simple to minutely bordered pits
- Fibres thin-walled
- Paratracheal axial parenchyma present
- Axial parenchyma also associated with ray cells
- Axial parenchyma in narrow bands
- Over 8 cells per parenchyma strand
- Minute alternate pitting in parenchyma cells: <= 4µm
- Rays 1–5 cells wide
- Ray height: >1mm
- Rays with procumbent, square and upright cells mixed throughout the ray
- Perforation in ray cells
- 4–12/ rays per millimetre
- Very rare occurrence of prismatic crystals in ray cells

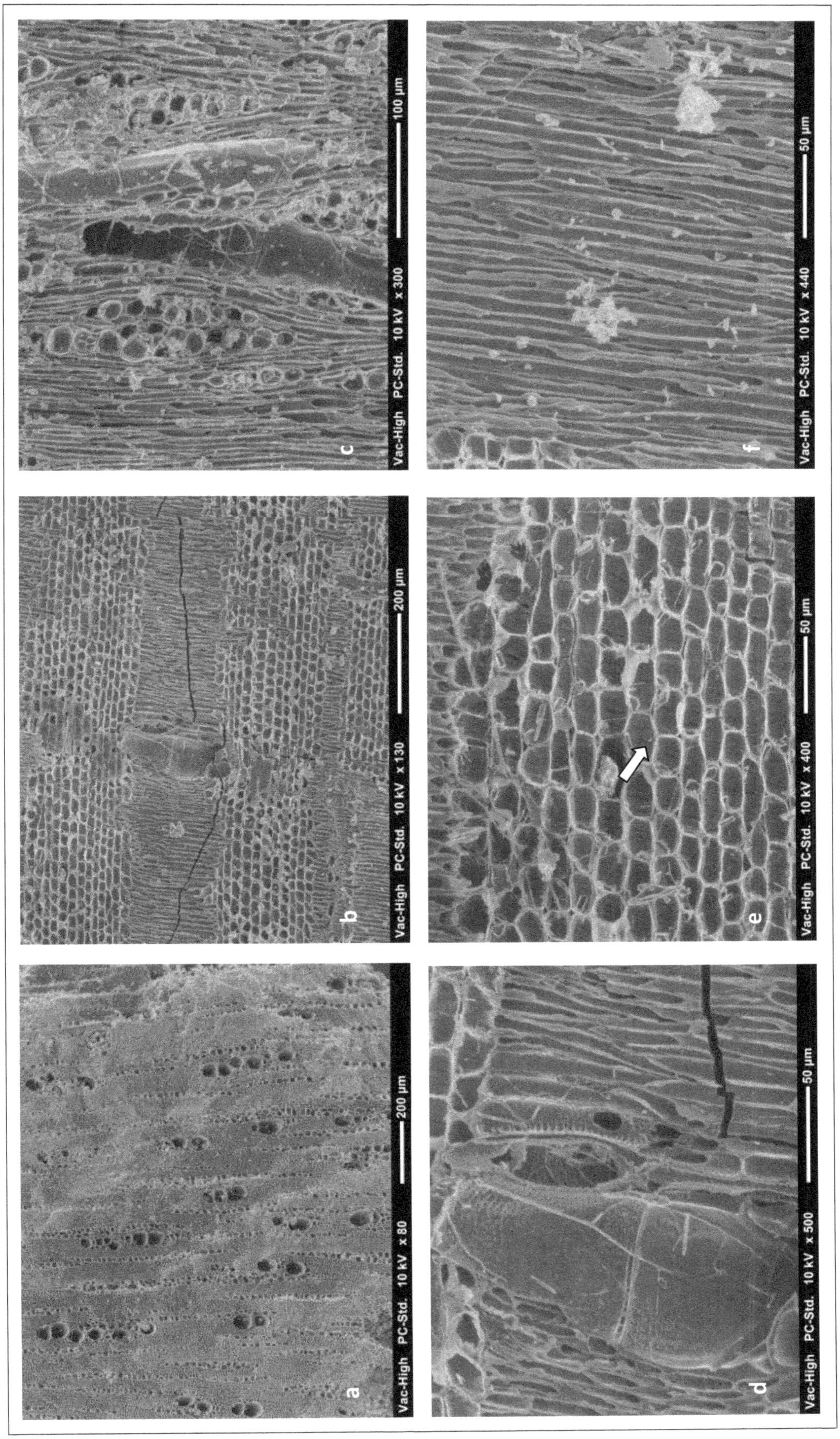

a: transversal; b and e: radial; c, d and f: tangential

Leopard Cave: CAPPARACEAE

IAWA Feature Code: 2,5,10,13,27,40,49,61,65,75,79,83,86,97,102,104,115,136,181

- Growth ring boundaries indistinct or absent
- Wood semi-ring porous
- Vessels in radial multiples of 4 or more common
- Solitary vessel outline rounded
- Simple perforation plates
- Alternate intervessel pitting
- Large intervessel pits: >= 10µm
- Mean tangential diameter of vessel lumina: <=50
- 40–100 vessels per square millimetre
- Gum and other deposits in vessels and ray cells
- Helical thickenings in ground tissue fibres (e)
- Septate fibres present (d)
- Fibres thin-walled
- Axial parenchyma absent or extremely rare
- Minute alternate pitting in parenchyma cells: <= 4µm
- Rays 1–3 cells wide
- Ray height: >1mm
- Rays with procumbent, square and upright cells mixed throughout the ray
- 4–12/ rays per millimetre
- Tension wood present

a: transversal; b and e: radial; c, d and f: tangential

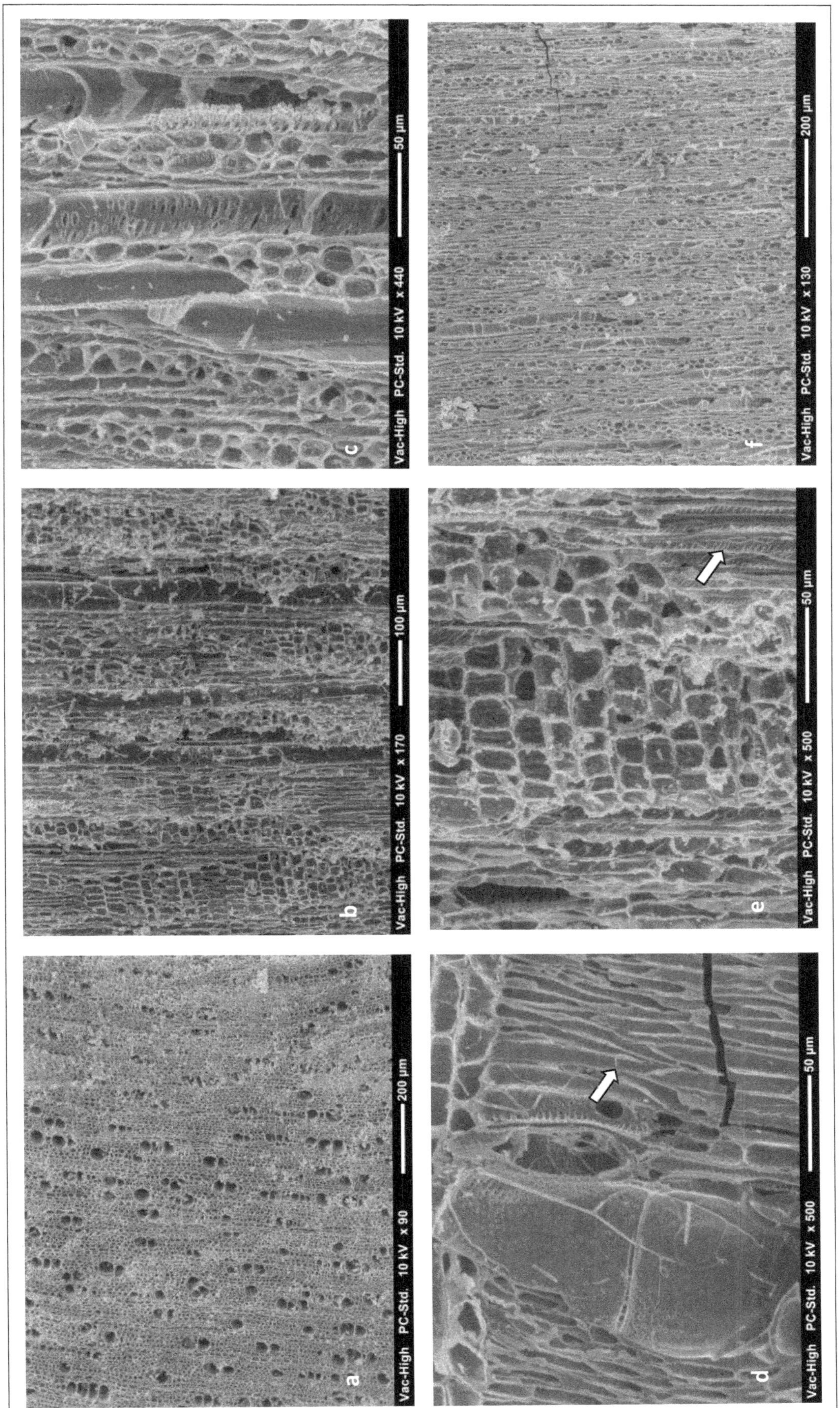

Leopard Cave: CAPPARACEAE - *Boscia* type

IAWA Feature Code: 2,5,8,11,13,22,24,40,49,60,61,65,75,76,86,97,102,104, 115,136,181

- Growth ring boundaries indistinct or absent
- Wood diffuse porous
- Vessels in dendritic pattern
- Vessels clusters common
- Simple perforation plates
- Alternate intervessel pitting
- Minute intervessel pits: <= 4µm
- Mean tangential diameter of vessel lumina: <=50µm
- Vessels of two distinct diameter classes
- 40–100 vessels per square millimetre
- Vascular tracheids present
- Helical thickenings in ground tissue fibres
- Fibres thin-walled
- Paratracheal axial parenchyma present
- Axial parenchyma also diffuse
- 5–8 cells per parenchyma strand
- Minute alternate pitting in parenchyma cells: <= 4µm
- Rays 1–3 cells wide
- Ray height: >1mm
- All rays procumbent
- Perforated ray cells (d and e)
- 4–12/ rays per millimetre
- Radial canals present, with up to 12 boarding cells (c)
- Radial canals with a domed appearance
- Alternate minute interlay pitting
- Tension wood present (e)

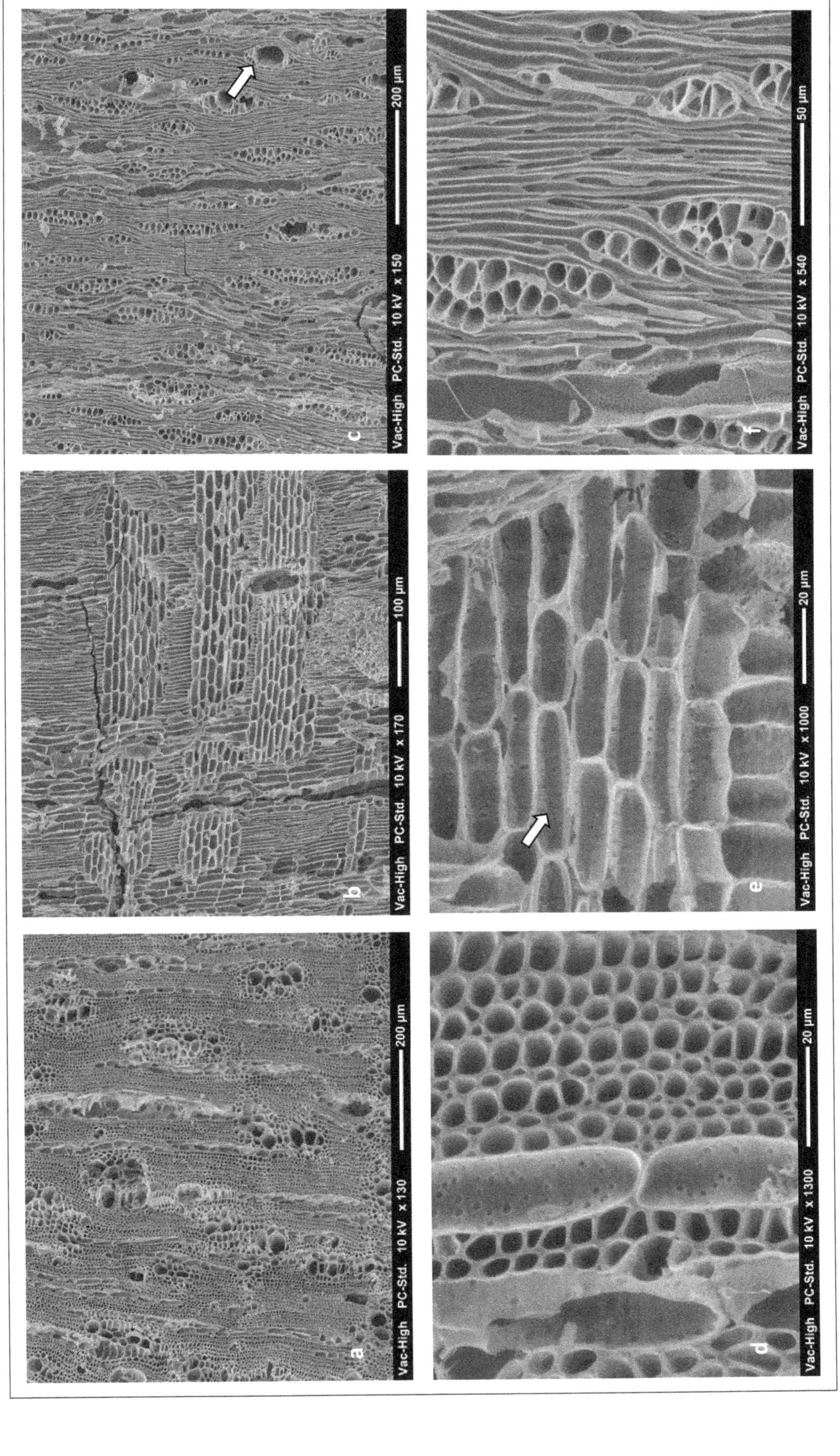

a: transversal; b and e: radial; c, d and f: tangential

Leopard Cave: COMBRETACEAE - *Terminalia prunoides* type

IAWA Feature Code: 2,4,9,10,13,22,24,41,50,60,61,65,75,76,86,96,97,102,115, 136,142,181

- Growth ring boundaries indistinct or absent
- Wood semi-ring porous
- Vessels mostly solitary and rarely in pairs or radial multiples
- Solitary vessel outline rounded
- Simple perforation plates
- Alternate intervessel pitting
- Minute intervessel pits: <= 4µm
- Mean tangential diameter of vessel lumina: 50–100µm
- Gums and other deposits in vessels
- >=100 vessels per square millimetre
- Fibres thin to thick-walled
- Axial parenchyma diffuse
- Minute alternate pitting in parenchyma cells: <= 4µm
- Rays 1–2 cells wide (b and c)
- Ray height: >1mm
- Rays with procumbent, square and upright cells mixed throughout the ray
- 4–12/ rays per millimetre
- Gums and other deposits in ray cells
- Gums and other deposits in vessels (a)
- Druses in ray parenchyma cells
- Crystal idioblasts present in ray-parenchyma cells (e)
- Prismatic crystals in chambered axial parenchyma cells
- Tension wood in ground tissue fibres

a and d: transversal; b and e: radial; c and f: tangential

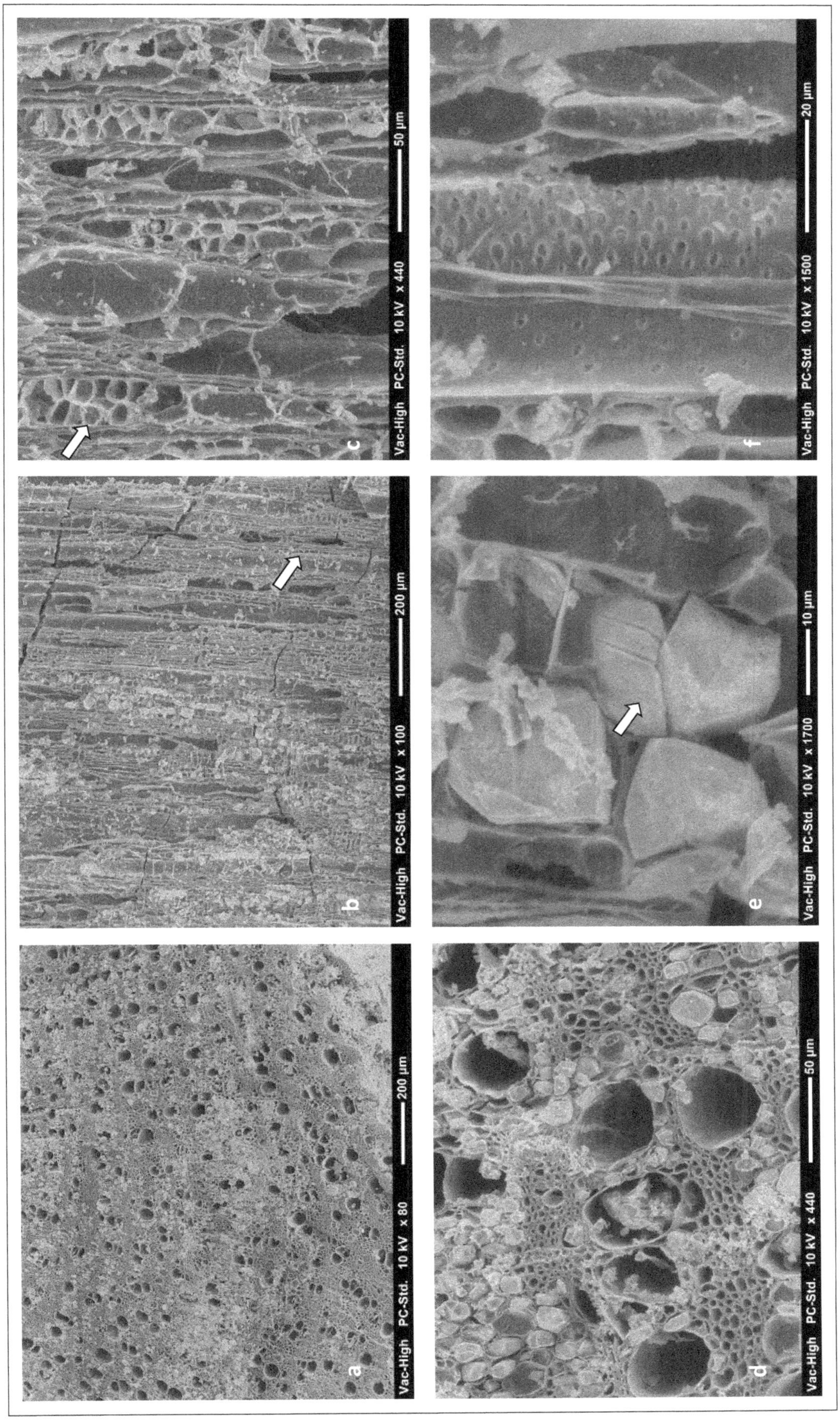

Leopard Cave: COMBRETACEAE - *Combretum molle*

IAWA Feature Code: 2,5,9,13,22,24,41,48,65,75,83,96,102,106,115,136,181

- Growth ring boundaries indistinct or absent
- Wood diffuse-porous
- Vessels mostly solitary (more than 90%)
- Solitary vessel outline rounded
- Simple perforation plates
- Alternate intervessel pitting (f)
- Minute intervessel pits: <= 4μm
- Mean tangential diameter of vessel lumina: 50–100μm
- 20–40 vessels per square millimetre
- Septate fibres present
- Fibres thin-to thick- walled
- Paratracheal axial parenchyma present
- Axial parenchyma confluent
- 5–8 cells per parenchyma strand
- Minute alternate pitting in parenchyma cells: <= 4μm
- Rays exclusively uniseriate
- Cupresoid shaped ray cell (d)
- Ray height: >1mm
- Body ray cells procumbent with one row upright and/or square marginal cells
- 4–12/ rays per millimetre
- Crystal idioblasts and silica body present in vessel-ray cells (e)
- Diffuse included phloem present (a)
- Druses in axial parenchyma cells

a: transversal; b and e: radial; c, d and f: tangential

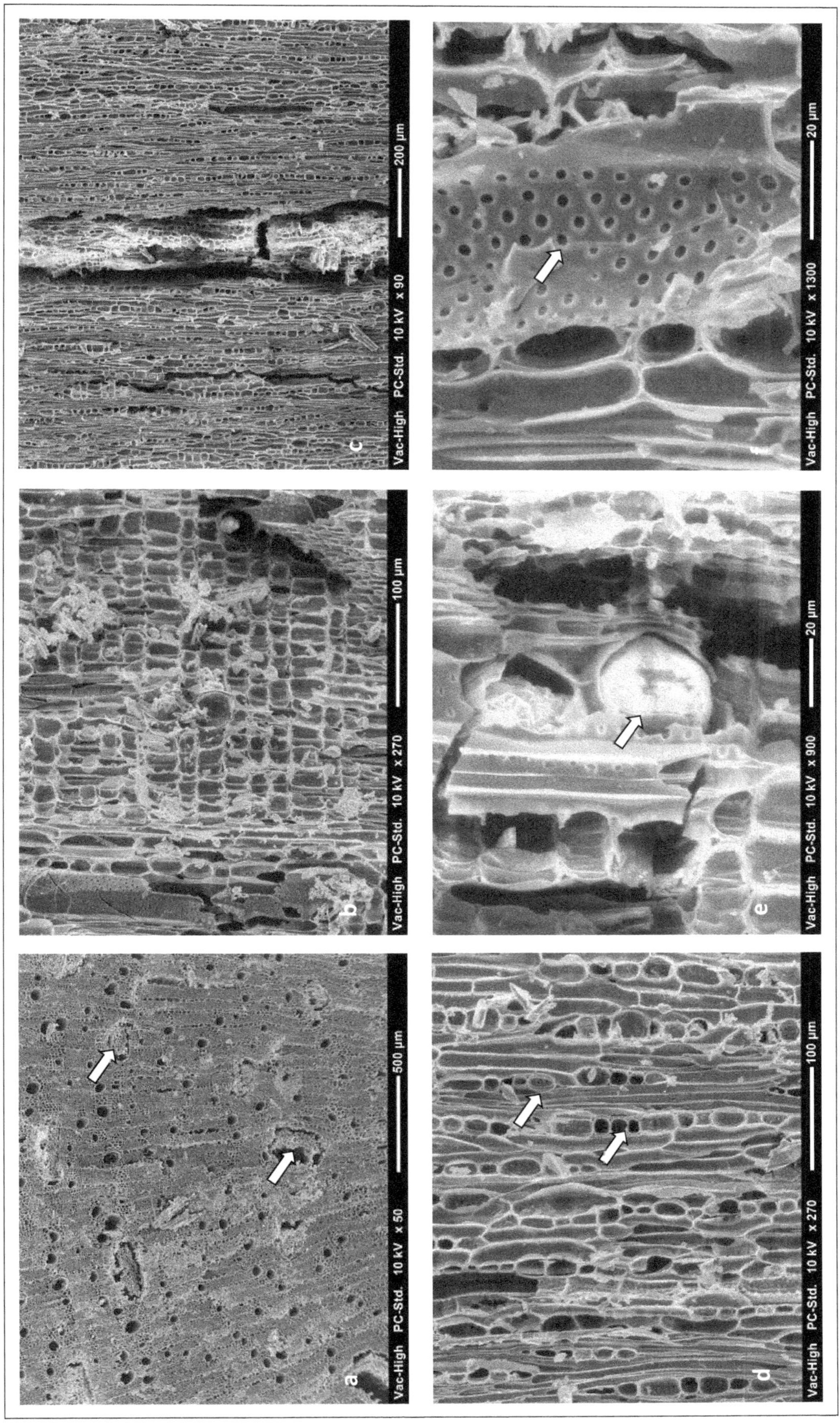

Leopard Cave: COMBRETACEAE - *Terminalia* type

IAWA Feature Code: 2,5,9,10,13,22,24,41,49,65,75,78,83,96,97,102,106,115, 136,156,181

- Growth ring boundaries indistinct or absent
- Wood diffuse-porous
- Vessels mostly solitary (more than 90%) and rarely in pairs or radial multiples
- Solitary vessel outline rounded
- Simple perforation plates
- Alternate intervessel pitting
- Minute intervessel pits: <= 4µm
- Vestured pits
- Vessels of two distinct diameter classes present
- Mean tangential diameter of vessel lumina: 50–100µm
- Tyloses present
- Vasicentric tracheids present
- 40–100 vessels per square millimetre
- Septate fibres present
- Fibres thin-to thick- walled
- Axial parenchyma scanty paratracheal
- Diffuse and banded axial parenchyma also present
- 5–8 cells per parenchyma strand
- Minute alternate pitting in parenchyma cells: <= 4µm
- Rays 1–2 cells wide
- Ray height: >1mm
- Body ray cells procumbent with one row upright and/or square marginal cells
- 4–12/ rays per millimetre
- Crystal idioblasts present in vessel-ray cells
- Prismatic crystals in chambered axial parenchyma cells
- Prismatic crystals in enlarged ray cells
- Styloids/Elongated crystals (c)
- Silica bodies present (d)

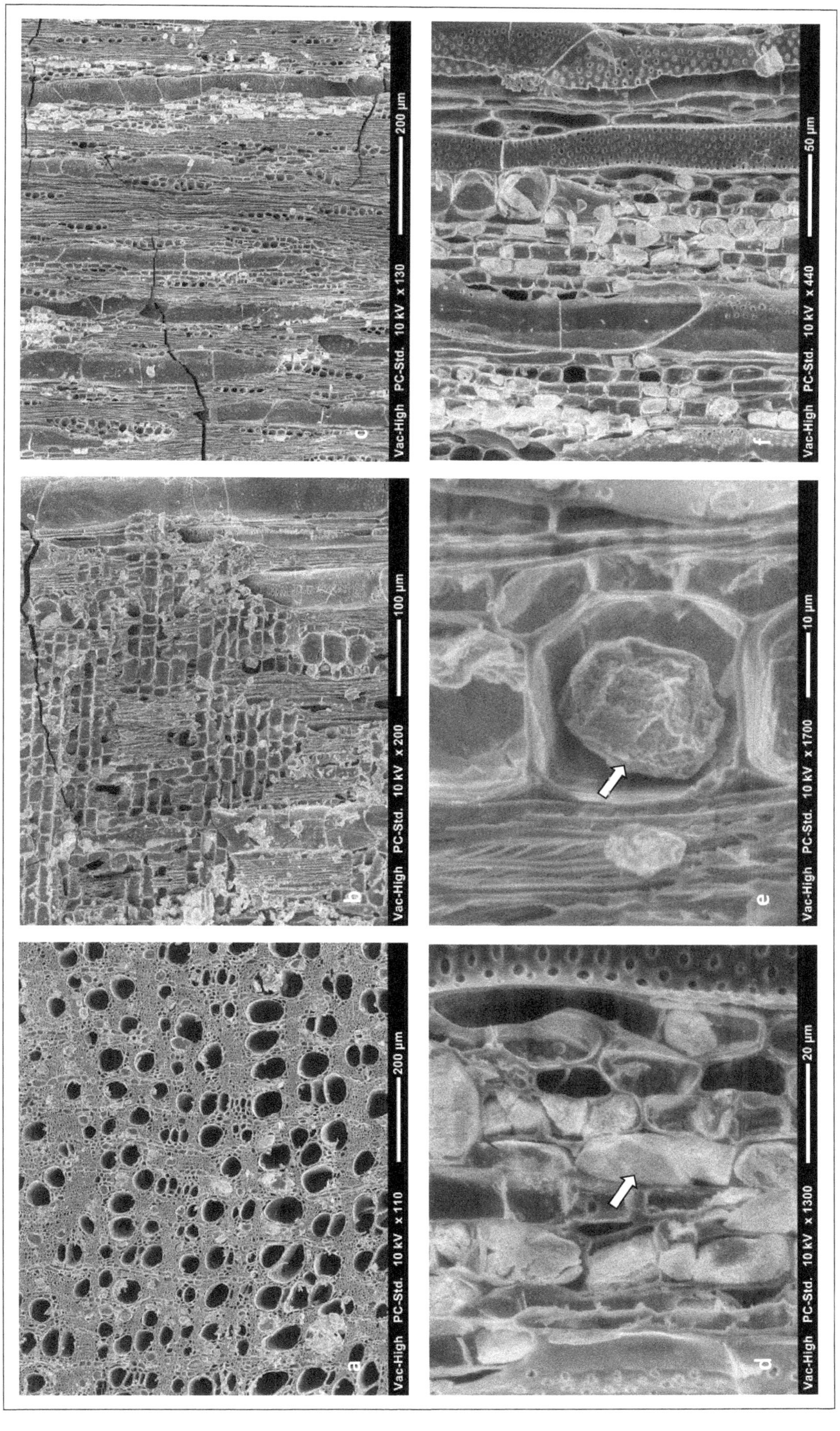

a: transversal; b and e: radial; c, d and f: tangential

Leopard Cave: ANACARDIACEAE- *Sclerocarya birrea*

IAWA Feature Code: 2,5,9,13,22,24,42,47,75,79,83,96,98,102,104,115,136,156,181

- Growth ring boundaries indistinct or absent
- Wood semi-ring-porous
- Vessels exclusively solitary (90% or more)
- Simple perforation plates
- Alternate intervessel pits
- Large intervessel pits - >=10 µm
- Mean tangential diameter of vessel lumina: 50–100 µm
- 20–40 vessels per square millimetre
- Vascular/vasicentric tracheids present
- Fibres with simple to minutely bordered pits
- Septate fibres present
- Axial parenchyma vasicentric
- Axial parenchyma bands more than three cells wide
- Ray width 1 to 3 cells
- Ray height > 1 mm
- All ray cells procumbent
- >= 12 /mm
- Prismatic crystals present, in ray cells and in fibre cells
- Crystals in enlarged cells
- Southern Africa (south of the Tropic of Capricorn) (Brazier and Franklin region 79)

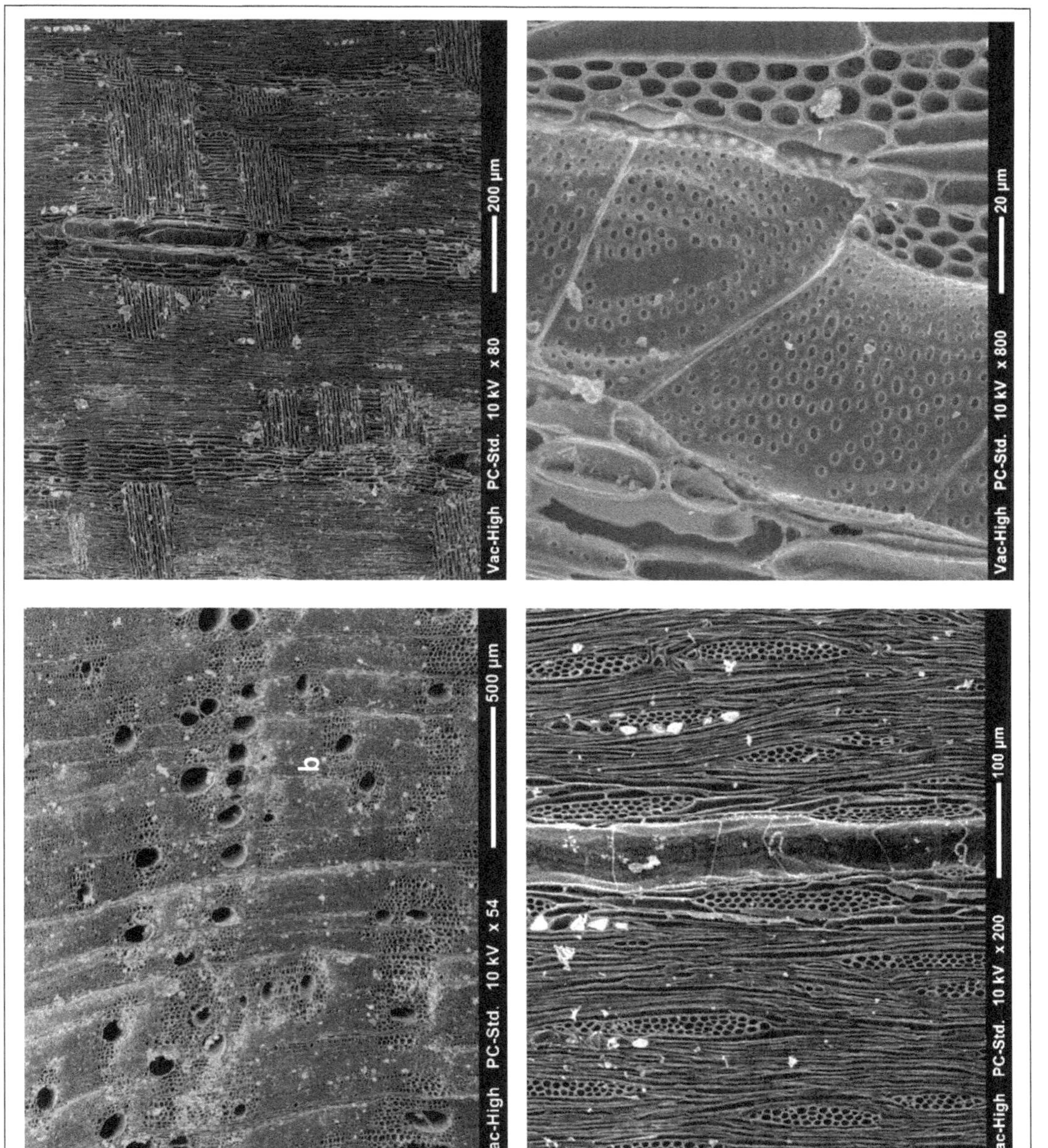
Vac-High PC-Std. 10 kV x 80
200 µm
Vac-High PC-Std. 10 kV x 800
20 µm
b
Vac-High PC-Std. 10 kV x 54
500 µm
Vac-High PC-Std. 10 kV x 200
100 µm

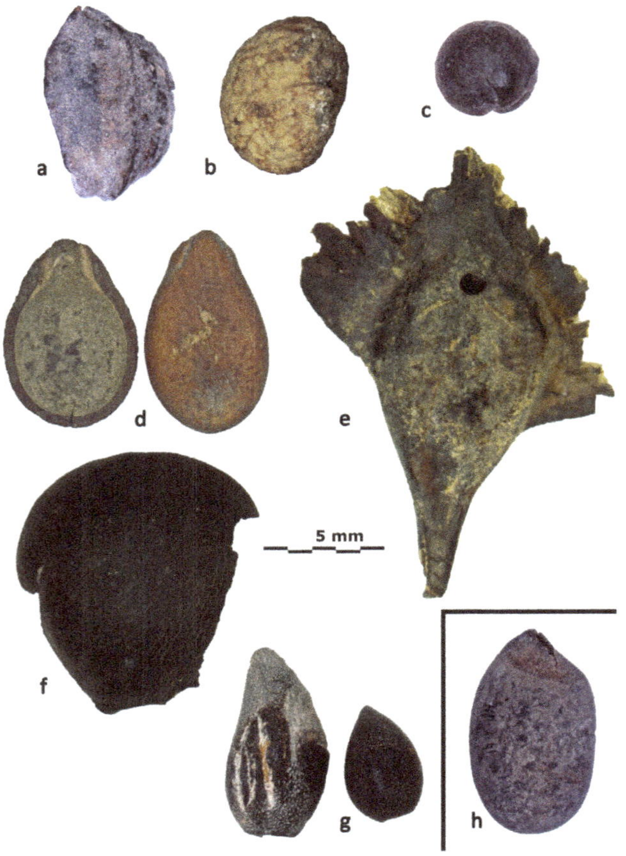

Figure 57. Examples of archaeological seeds recovered from Geduld a) *Commiphora sp,* b) *Boscia sp,* c) *Euclea pseudebens,* d) *Citrullus lanatus/erichorrus,* e) *Terminalia type,* f) *Tylosema esculentum,* g) *Lapeirousia sp,* h) *Ricinus communis.*

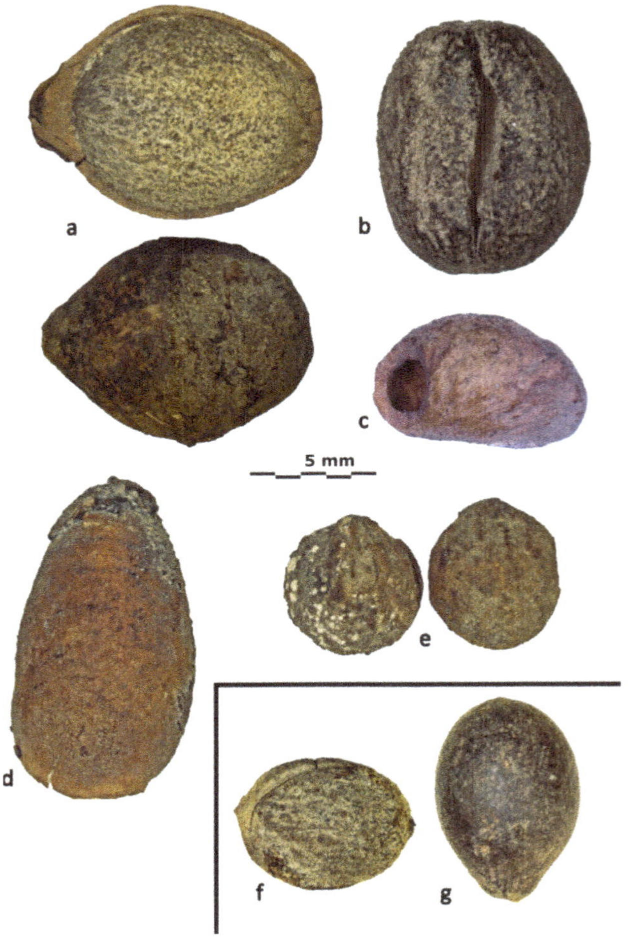

Figure 58. Examples of archaeological seeds recovered from Geduld and Leopard Cave a) Split *Ximenia americana/caffra endocarp,* b) *Ximenia americana/caffra kernel,* c) *Vangueria infausta,* d) *Berchemia discolour,* e) *Grewia spp,* f) *Ziziphus mucronata,* g) *whole Ximenia americana/caffra endocarp.*

5.2. Leopard Cave: results of the anthracological study

The analysis of a total of 1955 charcoal fragments selected from 19 samples collected during the excavation at Leopard Cave in 2012-15 (see Materials and methods) has identified ten botanical taxa belonging to four different plant families: Anacardiaceae, Capparaceae, Combretaceae and Fabaceae (Table 19, Table 20). The analysed fragments belong to two chronological phases identified on the basis of the stratigraphy and C14 dating of the excavated levels. The earlier Phase 1 (3700±150 - 3200±150 cal BP) has yielded fewer charcoal pieces (N=637) than the later Phase 2 (2600±150 -2100±100 cal BP) (N=1318). This difference in numbers is due mainly to the availability of more samples from the upper levels. Together the two phases cover a timespan of a little more than 1500 years.

All through the occupational sequence, wood from Vachellia/Senegalia trees (*Vachellia/Senegalia* spp.) dominates the charcoal record. Several species from this genus may be present, but it has not been possible to distinguish particular species for reasons that are explained in chapter 3. Besides representing the largest proportions of the identified charcoal fragments (43,3% in Phase 1, 49,6% in Phase 2), *Vachellia/Senegalia* wood is present in all the studied samples and seems thus to have constituted a common fuel at the Late Stone Age cave-site. In some cases, it was not possible to make a clear distinction between *Vachellia/Senegalia* -wood and that of *Dichrostachys*, another taxon from the Fabaceae family. When *Vachellia/Senegalias* and the category called *Vachellia/Senegalia/Dichrostachys* are considered together, species from the Fabaceae family correspond to between 63,3 percentof the total number of identified charcoal fragments in Phase 1 and 67,7 percent in Phase 2.

The second most frequently encountered group at Leopard Cave are the *Combretaceae* accounting for 22,9 percent of the charcoal fragments in Phase 1 and 26,2 percentin Phase 2. Within this family, species belonging to the *Terminalia* genus (*Terminalia prunioides* sp. *and Terminalia type*) are best represented, followed by those from the *Combretum* genus (*C. apiculatum* and *C. molle*). If considered together *Combretum* and *Terminalia* species are present in almost all of the studied contexts and the wood from

Table 19. Absolute numbers and frequencies of the different wood taxa identified from the occupation Phase I at Leopard Cave organised according to their layers and in alphabetical order (family followed by genus)

Phase		Phase 1							
Square		N7,O8,P7						Total N	%
Depth (cm)		186–190	190–196	196–200	206–210	Foyer_1–229	Foyer_2–259		
Anacardiaceae	*Sclerocarya birrea*							0	0.0%
Capparaceae	*Boscia type*	1		4	9	51		65	10.2%
	Maerua/Boscia					11	12	23	3.6%
Combretaceae	*Combretum apiculatum*	5	4	2				11	1.7%
	Combretum molle					5	32	37	5.8%
	Combretum spp.	4		3				7	1.1%
	Terminalia prunioides	4		7		13	7	31	4.9%
	Terminalia type				11	20	29	60	9.4%
Fabaceae	*Acacia spp.*	33	17	27	8	65	126	276	43.3%
	Acacia/Dichrostachys				16	73	38	127	19.9%
	Total	**46**	**21**	**39**	**35**	**176**	**232**	**637**	**100%**

the Combretaceae family thus also seems to have been frequently collected, together with that of *Vachellia/Senegalia* during the entire occupation of the site.

The Capparaceae family is represented by two taxa: Boscia sp. and *Boscia/Maerua* corresponding to 13,8 percent and 5,8 percent of the charcoal pieces in the Phases 1 and 2 respectively. Finally, the Anacardiaceae family is represented by one species, *Sclerocarya birrea*, absent in Phase 1 and attested by a low number of fragments (N=4) from one single layer in Phase 2 (135–140 cm). The presence of this species is intriguing not only because of its rarity in the Leopard Cave assemblage but also due to its complete absence from the surrounding environment today.

When comparing the representation of the different taxa according to the two phases (Figure 59) the similarities are striking. The same plant groups are dominant throughout the sequence and proportions are not significantly different from one phase to the other, except in the case of *Boscia* for which a certain decrease can be noted, from 10,2 percent in Phase 1 to 3,0 percentin Phase 2. It is also perceptible that *Sclerocarya birrea* is only present in Phase 2, represented by a low number of fragments (N=4). No major evolution of the vegetation formations around Leopard Cave can thus be detected through the charcoal study. It is possible that a multiplication of the number of charcoal fragments studied from the different layers could introduce some more detail on possible variations through time, but it is unlikely that any important change of the surrounding vegetation cover has taken place during the occupation of the cave site.

5.2.1. Results of the seed analysis

The seed/fruit assemblage from Leopard Cave consisted of both desiccated and carbonised remains most of which belong to woody species with only a few from herbaceous taxa being represented. The remains coming from trees and shrubs consist predominantly of whole or fragmented fruit stones (botanically endocarps) sometimes still enclosing the seed or the seeds, according to the species. Only the later Phase 2 is represented among this category of remains.

The analysis of 233 relatively well-preserved items has allowed the identification of 9 different taxa as well as a tenth type that has not yet been identified (Table 21). These taxa represent as many botanical families as possible: Burseraceae, Combretaceae, Cucurbitaceae, Euphorbiaceae, Fabaceae, Olacaceae, Poaceae, Rubiaceae and Tiliaceae. The genus *Grewia* is largely predominant among these remains, representing 53 percent of the total number. *Vangueria infausta* (17 percent) is also common in the samples while other taxa occur less frequently and in lower proportions. A concentration of grass seeds is recorded from the layer excavated at 130-135 cm depth, but otherwise seeds from annual plants are rare. The grass remains could not be identified to the genus or species level but could correspond to one of the varieties of grasses found in the surroundings. However, none of those is described as edible, contrary to the case of *Sporobolus* cited during the ethnobotanical study conducted among the villagers of Tubusis.

5.2.2. First observations on the macrobotanical record at Leopard Cave

A significant observation occurs when comparing the charcoal and seed/fruit remains from Leopard Cave, i.e. very little taxonomic overlap between the two categories. Certainly, the taxa identified by the anthracological study are generally absent from the seed/fruit assemblage and vice-versa. The only taxa present in both are *Terminalia* sp. (Possibly *T. prunioides*) and Fabaceae, which may include

Table 20. Absolute numbers and frequencies of the different wood taxa identified from the occupation Phase IIat Leopard Cave organised according to the depth and in alphabetical order (family followed by genus)

Phase		Phase 2														
Square		M6, N6, N7, O7													Total N	%
Depth (cm)		0–10	124–130	130–135	135–140	136–146	140–145	145–150	146–156	155–160	160–165	165–170	170–182	176–186		
Anacardiaceae	*Sclerocarya birrea*				4										4	0.3%
Capparaceae	*Boscia* type		7		4		3	2	4	1	11	3	3	1	39	3.0%
	Maerua/Boscia				2		11				14		11		38	2.9%
Combretaceae	*Combretum apiculatum*		1	1	1	1	4	4	2	28	3		4	1	50	3.8%
	Combretum molle									6	10	6			22	1.7%
	Combretum spp.	7	1	1	5	5	5	8	12	8	9	1	3		65	4.9%
	Terminalia prunioides	2	22	1	35	15	12	15	4	11	2		3	2	124	9.4%
	Terminalia type		1	7			11	8	4		27	2	24		84	6.4%
Fabaceae	*Acacia* spp.	6	102	73	57	39	43	58	33	74	80	16	46	27	654	49.6%
	Acacia/Dichrostachys		*3*		5	0	*56*	3	15	6	99	2	49		238	18.1%
	Total	**15**	**130**	**83**	**103**	**60**	**131**	**96**	**70**	**133**	**230**	**27**	**129**	**30**	**1318**	**100%**

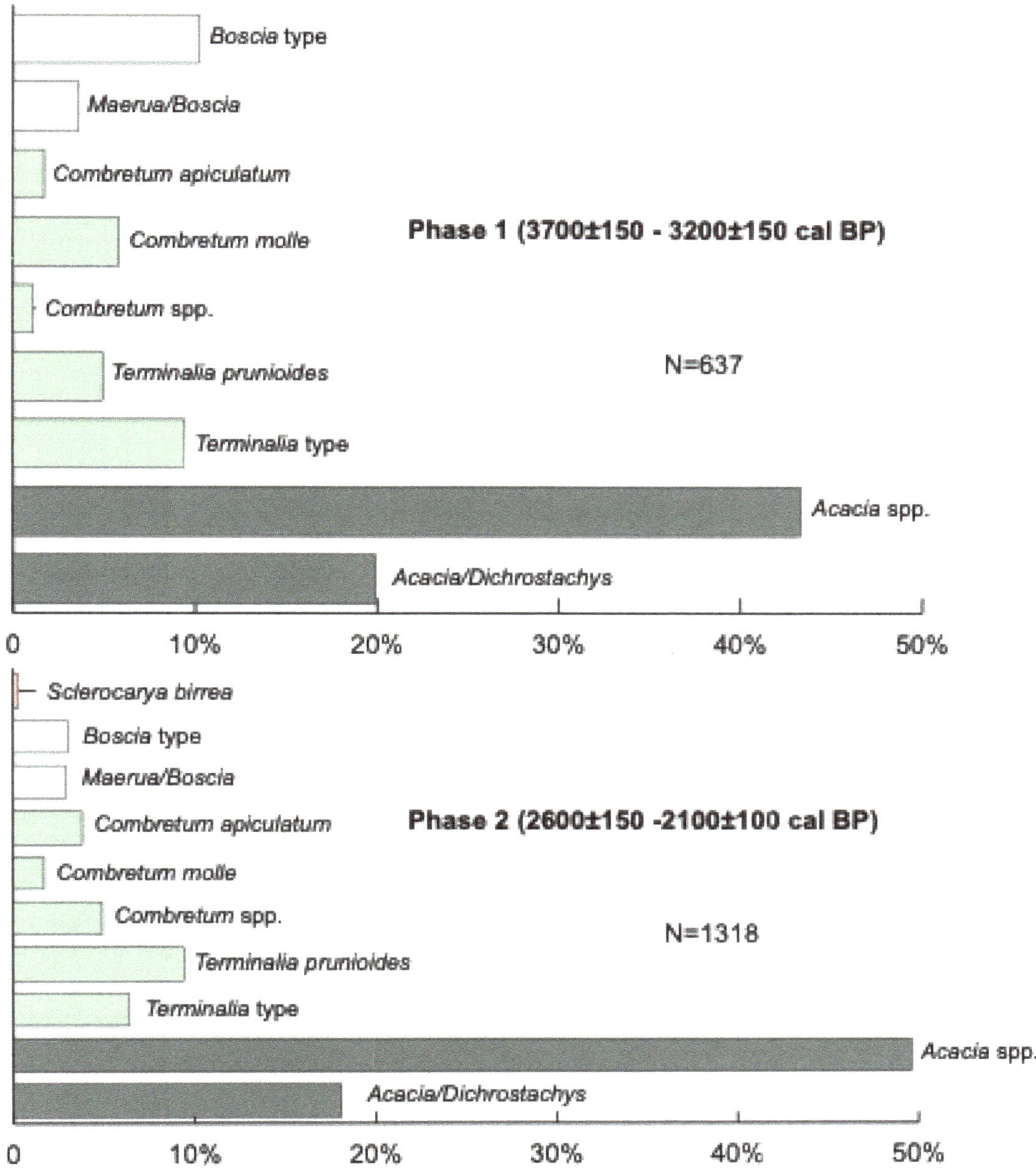

Figure 59. Charcoal diagrams from Leopard Cave showing relative proportions of the identified taxa in Phases 1 & 2.

a variety of woody trees and shrubs that are presently found in the region. However, their frequencies are very low, with only 1 and 3 percent of the total assemblage, respectively.

Between 86,2 percent (Phase 1) and 93,9 percent (Phase 2) of the charcoal fragments are identified as belonging to the Fabaceae and Combretaceae families. Neither produces fruits used as human food (cf the ethnobotanical study). Fruit-bearing species are more common in the Capparaceae family but are not attested in the seed record from Leopard Cave. 75 percent of the taxa identified by the seed/fruit analysis correspond to trees with edible fruit (*Vangueria infausta, Ximenia americana/caffra, Commiphora* sp., *Grewia* sp.*)*. The Tsamma melon (*Citrullus lanatus)* is a perennial creeper whose flesh and seeds are edible.

5.3. Geduld: results of the anthracological study

The analysis of 691 charcoal fragments contained in nine samples collected during the excavation of Geduld by the American team identified eight taxa belonging to three botanical families: Capparaceae, Combretaceae and Fabaceae (Table 22, Figure 60). No precise chronological phases have been determined and the charcoal results are here considered as a whole, even though radiocarbon dates from the site indicate an occupation spanning several centuries in the late 3rd and early 2nd millennia BP (Smith et al. 1995).

As at Leopard Cave, the majority of the charcoal from Geduld belongs to the Fabaceae family. Fragments identified as *Vachellia/Senegalia* spp., probably *Vachellia/Senegalia*

Table 21. Result of the seed/fruit analysis at Leopard Cave

	Phase	Phase 2											Total N	%
	Square	M6,N6,N7,O7												
	Depth (cm)	60–70	124–130	130–135	135–140	140–153	147–155	150–160	155–160	160–167	165–170	175–180		
Burseraceae	***Commiphora* sp.**			2		1							**3**	1.0%
Combretaceae	***Terminalia* sp.**				2								**2**	1.0%
Curcubitaceae	***Citrullus lanatus* sp.**						7	1					**8**	3.0%
Euphorbiaceae	***Ricinus communis***				3			2	3				**8**	3.0%
Fabaceae	**Fabaceae**					3					5		**8**	3.0%
Olacaceae	***Ximenia americana/caffra* sp.**		2			1			5				**8**	3.0%
Poaceae	**Poaceae**			20					1				**21**	9.0%
Rubiaceae	***Vangueria infausta* sp.**		12		3	1	7	3	13				**39**	17.0%
Tiliaceae	***Grewia* sp.**	20	5	3	41	3	5	26	3	9		8	**123**	53.0%
	Unid. 1				8	4				1			**13**	6.0%
	Total	**20**	**19**	**25**	**57**	**13**	**19**	**32**	**25**	**10**	**5**	**8**	**233**	**100%**

Table 22. Absolute numbers and frequencies of the different taxa identified at Geduld organised according to their square and level

Square		B6	B6	B6	B7	B7	B8	B8	B9	C7	Total N	%
Level		L7	L6	L5	L7	L5	L4	L3	L3	L13/L11		
C14 date layer (BP)		/	/	1790±80	/	/	/	/	1791±50*, 1676±25*	2300±50, 2110±60		
Caesalpinaceae	*Colophospermum mopane*	3									3	0.4%
Capparaceae	*Boscia/Maerua*	5	1		5			2		1	14	2.0%
	Capparaceae type	3	7		5			14	5		34	4.9%
Combretaceae	*Terminalia* type	54	7			4			3	18	86	12.4%
Fabaceae	*Acacia* spp.	9	21	2	3	23	1	4	4	38	105	15.2%
	Acacia/Dichrostachys	225	20		2	49		5	1	19	321	46.5%
	cf *Acacia*	71	3	1	8	7	4	18	7	3	122	17.7%
	Peltophorum africanum	4				1				1	6	0.9%
Total		371	59	3	23	84	5	43	20	80	691	100.0%

and *Vachellia/Senegalia/Dichrostachys* represent almost 80 percent of the total charcoal record. Two other taxa from the *Fabaceae* family are also attested: *Colophospermum mopane* and *Peltophorum africanum* occurring in low proportions (0,4 and 0,9 percentrespectively).

The Combretaceae family is the second most well-represented group, with what it seems to be *Terminalia* according to the anatomical characteristics observed. This genus accounts for 12,4 percentof the identified charcoal pieces.

The third group present at Geduld are the Capparaceae representing a total of 6,9 percent divided between *Boscia/Maerua* and Capparaceae type.

5.3.1. Results of the seed analysis

The site of Geduld delivered a large assemblage of seed/fruit remains (N=512) that are the subject of a study published in 1995 (Yates et al. 1995). Nevertheless, a choice was made to include this material here, in order to confirm and to elucidate previous identifications, add supplementary items and compare the results obtained with those of the charcoal analysis. The assemblage consists of a mixture of desiccated and carbonised remains, endocarps (whole and fragmentary), seeds and bulbs. Examination identified sixteen taxa belonging to twelve plant families: Burseraceae, Capparaceae, Combretaceae, Curcubitaceae, Cyperaceae, Ebenaceae, Euphorbiaceae, Fabaceae, Iridaceae, Olacaceae, Rhamnaceae and Tiliaceae (Table 23).

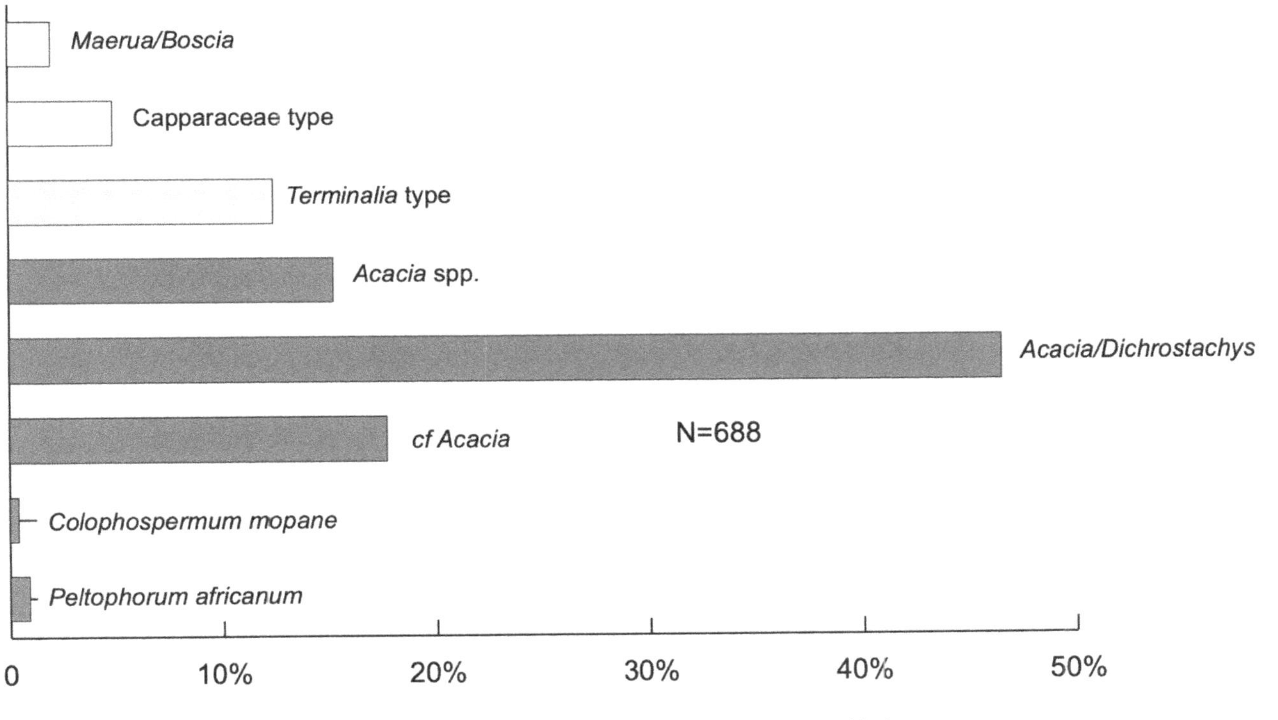

Figure 60. Charcoal diagram from Geduld showing relative proportions of the identified taxa.

Table 23. Result of the seed/fruit analysis at Geduld

Square		B5-6	B6	B6	B7	B8	B9	C6	C7	C7	Total	%
Level		L1/4	L5	L7	L6/L7	L1/4	L1/3	L8	L8/L10	L11/14		
Square and level		Bedding	Ash/bedding	Ash/ bedding	Ash/ bedding	Dung/Ash/ Hearth	Dung/Ash/ Burrow	Bedding/ Ash	Ash	Sand/Grass		
C14 date layer (BP)		/	1790±80	/	/	/	1791±50*, 1676±25*	/	/	2300±50, 2110±60		
Burseraceae	*Commiphora* spp.			1			1	1	2		5	1.0%
Capparaceae	*Boscia albitrunca*				1						1	0.1%
	Boscia foetida				83		1				84	16.0%
Combretaceae	*Terminalia* sp.			3							3	1.0%
Curcubitaceae	*Citrullus lanatus*	14	7	46	8	6	34	1	16	5	137	27.0%
Cyperaceae	*Cyperus* sp.			3		4	3		4	4	18	4.0%
Ebanaceae	*Euclea pseudebenus*	2		4	2	2	8	1	1		20	4.0%
Euphobiaceae	*Ricinus communis*			1							1	0.1%
Fabaceae	*Acacia* sp.	2					4				6	1.0%
	Tylosema esculentum	1	3	4	2	4	10	12	6		42	8.0%
	Elephantorrhiza elephantina								2		2	0.4%
Iridaceae	*Lapeirousia* sp.					3	1	1			5	1.0%
Olacaceae	*Ximenia americana/caffra*	1	2	9	7	10	5	4	13	5	56	11.0%
Rhamnaceae	*Berchemia discolor*	1	1	2		7	10		1		22	4.0%
	Ziziphus mucronata	5		2	1	5	14		3	1	31	6.0%
Tiliaceae	*Grewia* spp.	11	3	5		7	26	3	2	1	58	11.0%
	Unid		1			2	16		1	1	21	4.0%
	Total	**37**	**17**	**80**	**104**	**50**	**133**	**23**	**51**	**17**	**512**	**100%**

Table 24. Absolute numbers and frequencies of the different taxa identified at Toteng organised according to their square and level

ID_N°		T9_10	T8_9	T7_8	T6_7	T5_6	T4_5	T3_4	T2_3	T1_2	1	Total N	%
Depth (cm)		**100–110 cm**	**90–100 cm**	**80–90 cm**	**70–80 cm**	**60–70 cm**	**50–60 cm**	**40–50 cm**	**30–40 cm**	**20–30 cm**	**20 cm**		
Capparaceae	Capparaceae		5									5	2.5%
Combretaceae	*Terminalia prunioides/ sirecea*	20		10			7	6	4	9		56	28.3%
Fabaceae	*Acacia* spp.	11	6	5	8	6	9	10	26	17	7	105	53.0%
	Acacia/ Dichrostachys		7	6	2	5				12		32	16.2%
Total		**31**	**18**	**21**	**10**	**11**	**16**	**16**	**30**	**38**	**7**	**198**	**100%**

Among these taxa, trees and shrubs are present as well as herbs and creepers, such as Tsamma melon (*Citrullus lanatus).*

The latter is present in all of the studied samples and accounts for the highest number of seeds, representing more than on quarter of the total assemblage. Another ubiquitous species is the yellow plum (*Ximenia americana/ caffra)* from the Olacaceae family. It is represented by 11 percent of the identified seeds. A slightly higher number of remains are identified as *Boscia foetida* but they all come from one sample (L6/L7) except for one supplementary solitary seed. *Grewia, Tylosema esculentum* and *Ziziphus mucronata* represent 11,8 and 6 percent of the remains, respectively. Seeds/fruits from other taxa are less numerous, with a relative number varying between 4 and 0,1 Percent. Several species are present in the form of only one or two remains *(Boscia albitrunca, Ricinus communis, Elephantorrhiza elephantina).* The bulbs belonging to *Lapeirousia* sp. of the Iridaceae family constitute an interesting find that will be discussed further in chapter 6.

5.3.2. First observations on the macrobotanical record at Geduld

Several of the ligneous genera identified by charcoal at Geduld are also present in the seed/fruit assemblage (*Boscia, Terminalia*). Still, they are not identified with the same level of precision and might not belong to the same species. *Vachellia/Senegalia* and *Vachellia/Senegalia* -related taxa are predominant in the charcoal record but rare in the seed/fruit assemblage. This is a similar situation to Leopard Cave where no seeds/pods of *Vachellia/ Senegalia* have been found.

In general, the seed/fruit assemblage at Geduld Cave is diversified with sixteen taxa identified among which ten are identified at the species level. Literature shows that the majority, if not all, of the identified plants, are still available in the surroundings of Geduld with *Tylosema esculentum* from the *Fabaceae* family constituting an important part of the diet in many parts of Southern Africa.

5.4. Toteng: results of the anthracological study

Analysis of 198 fragments contained in 10 samples collected from the stratigraphic sounding opened at Toteng in 2016 has identified four taxa belonging to three different plant families: Capparaceae, Combretaceae and Fabaceae (Table 24). As in the two previously described Namibian sites, at Toteng the predominant taxa in the samples is *Vachellia/Senegalia* spp. present in all levels and representing more than 50 percent of the identified fragments. If added, the charcoal fragments identified either as *Vachellia/Senegalia* or *Dichrostachys,* the *Mimosoideae* subfamily of the Fabaceae represents nearly 70 percent of the analysed charcoal fragments.

The Combretaceae, represented by *Terminalia prunioides/ sericea,* appears in 60 percent of the samples and corresponds to 28, 3 percent of the total number of identified charcoal fragments. A fourth taxon identified as belonging to the Capparaceae family appears only in the level between 90 and 100 cm (Figure 61).

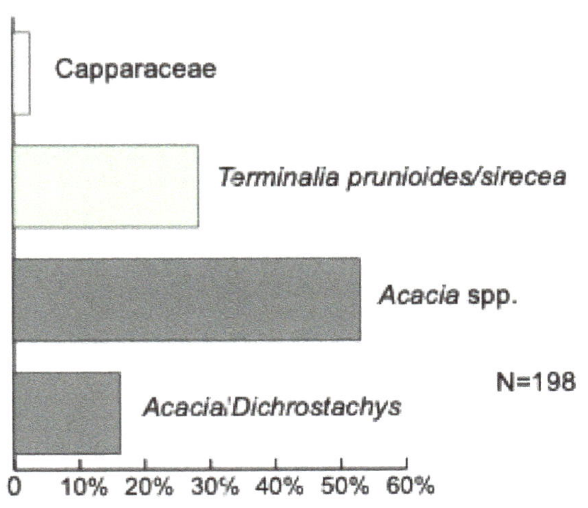

Figure 61. Charcoal diagram from Toteng showing relative proportions of the identified taxa.

5.5. Preliminary conclusions

Despite the geographical disparity of the three sites, and the unavailability of carpological material from the site of Toteng, the results from the charcoal and the seeds analysis show a uniform pattern. Woody taxa especially from the Fabaceae, Combretaceae and Capparaceae families and notably *Vachellia/Senegalia, Terminalia, Boscia and Combretum* species are dominant. Also present in all is a substantial amount of edible fruit, some belonging to taxa attested in the form of charcoal and seeds. The charcoal and seed assemblages from Leopard Cave and Geduld, show different taxa that are exclusive to each, with only a few appearing in both.

6

Discussion

6.1. Leopard Cave's arboreal species from charcoal

Leopard Cave alone of the three sites has a chronology that can for the moment be divided into two distinct phases: Phase 1 (3700±150–3200±150 BP) and Phase 2 (2600±150–2100±100 BP). However, the results do not allow for a detection of any change in the vegetation cover between the two occupational phases as shown in very similar plant taxa. A very slight difference exists in the reduced presence of *Sclerocarya birrea* in Phase 2. Today this species is a component of the *Mixed bushveld* plant community found in most of the Southern African plateau.

The main vegetation community represented by the taxa identified at Leopard Cave is what Palmer and Staden (2006) describe as the *Arid Sweet Bushveld* dominated by several *Vachellia/Senegalia* species ranging from shrubs to trees growing up to more than 20 m. These formations also include *Vachellias/Senegalia* adapted to riparian[8] woodlands that develop mostly on deep alluvial soils. The predominant species in these formations is *Vachellia karoo*. This *Vachellia/Senegalia* community is accompanied by dwarf shrubs such as *Boscia foetida, Dichrostachys cinerea* and several shrubs from the genus *Grewia*. Today in the vicinity of the site, *Senegalia mellifera, A. tortilis, A. erubescens, A. robusta* and *Dichrostachys cinerea* grow among other species in the open plains while *A. karoo*, in association with other *Vachellia/Senegalia* shrubs, is observed in the valley and along dry streams in the Erongo escarpments as well as around the Omaruru River near the town of Omaruru. Wadley (1979) also notes the presence of "giant Acacias" accompanied by tamarisk trees and grasses around the site of Big Elephant Shelter (Erongo mountains) and the Erongo region in general. The *Vachellia/Senegalia /Dichrostachys* charcoal fragments from the archaeological record are characterised by broad rays (3–5 seriate) and mostly solitary vessels with a relatively large vessel lumina.

Today one of the taxa identified from Leopard Cave belong mainly to the *Mixed bushveld* plant community, composed of woody *Combretum* and *Terminalia* species. This plant community could be further classified as a *Combretum-Terminalia* woodland composed of *Combretum apiculatum* and *Combretum molle* as well as *Boscia albitrunca*. Even though these plant communities are considered as two separate entities, it is difficult to draw a boundary between them as the dominant species in each overlap and merge (Thackeray 1979). The two are therefore sometimes grouped as bush/shrub and open tree steppe (Wadley 1979).

The wood samples attributed to the Capparaceae family at Leopard Cave present a wide range of characteristics from vessels in radial multiples to thick but short rays (1–2 seriate) in *Boscia/Maerua* species, and a nested vessel pattern with radial canals in uni-biseriate rays in the *Boscia* type. The third type attributed to the Capparaceae family has vessels in radial multiples, with little or no axial parenchyma and differs from the *Boscia/Maerua* group by the absence of prismatic crystals. Today *Boscia albitrunca* occurs extensively in the open plains and on the *kopje* slopes around Leopard Cave and the neighbouring site of Big Elephant. *Boscia foetida* occurs in the open grassland and in the valley but is absent from the mountain slopes.

Concerning the charcoal identified as belonging to the Combretaceae family, *Combretum molle* and the *Terminalia* type both have solitary vessels with very small vessel lumina and narrow rays (1–2 seriate). *Combretum molle* also shows the typical characteristics of the Combretaceae family i.e. included phloem. *Terminalia prunioides* is characterised by vessels arranged in radial multiples with larger vessel lumina than the previous two taxa. All the plants identified as belonging to the Combretaceae family can be seen around the site today even though the only *Terminalia* species present is *T. prunioides*. Ecologically speaking, another *Terminalia species*, *T. sericea*, might have grown in the area in the past but if such was the case it seems to have disappeared some while ago as it is not mentioned from the neighbouring sites of Big Elephant and Fäckeltrager nor in the study of plant uses in dry areas of north-western Namibia (Sullivan 1998, 1999).

The Marula tree (*Sclerocarya birrea*) is present in very low numbers (0.3 percent of the total charcoal assemblage). It also belongs to the woodland plant group but requires a more humid environment than in the present-day arid Erongo (Cheikhyoussef and Embashu 2013). The wood of the species is recognisable by its large vessels and wide 2–4 seriate rays.

6.2. Arboreal species from seed analysis

The results of the seed/fruit analysis at Leopard Cave allows us to reconstruct the two same vegetation communities – the Mixed Bushveld (Figure 62) and the Arid Sweet Bushveld (Figure 63) – like those identified by the charcoal analysis. These are dominated by woody taxa such as *Grewia*, *Commiphora* and *Terminalia*. The

[8] The interface between land and a river or stream mainly occurring in arid and semiarid regions, the presence of surface water and groundwater fosters the development of distinct riparian zones replete with herbs, shrubs and trees (Lake et al. 2017).

Figure 62. Mixed bushveld (Southeastern Botswana) Photo: M. Mvimi.

Figure 63. Thorny thicket corresponding to the Sweet Arid bushveld (Southeastern Botswana) Photo: M. Mvimi).

presence of *Vangueria infausta* further indicates dry-grasslands (Daemane et al. 2010). This species together with *Commiphora* species is today associated with rocky midslopes and mountain ridges. Today, the genera *Ximenia,* the last of the woody taxa found at Leopard Cave, grows in association with dry *Vachellia/Senegalia* woodlands, wooded grasslands as well as rocky outcrops.

The seed assemblage further comprises several annual plants such as *Citrullus lanatus* and *Ricinus communis* as well as undetermined seeds from the Poaceae and Fabaceae families. *Ricinus communis* is a soft-woody shrub or small tree that can attain up to 6 m in height. It grows in riparian or stream-bank communities along seasonally dry streams or in open savannah. The glabrous seeds are about 2 cm long and have dark and light brown patterns. *Citrullus lanatus* is an herbaceous annual plant that grows in the grassland and bushland plant communities of Southern Africa, mostly on sandy soils. *C. ecirrhosus* is restricted to the Kgalagadi desert, the Namib and in the Richtersveld of the Northern Cape. These plants were not observed during the fieldwork periods in July-September as they develop after the summer rains and/or are restricted to the more arid Namib to the southwest. The seeds are ovate-elliptic in shape with a smooth endocarp that is sometimes patterned.

The Poaceae seeds could denote a variety of grasses growing throughout the open Savannah. One among other edible grass taxa consumed in the region today is *Sporobolus* (see chapter 4). In general, few grass remains are found in Leopard Cave. In addition to the woody *Fabaceae* species, herbaceous species from this family are also present. While they could not be identified at a precise level, several *Fabaceae* species are present in the above-described vegetation units.

6.3. Geduld's arboreal species from charcoal

Very similar vegetation patterns as observed for Leopard Cave can be reconstructed for Geduld. There is only a slight change concerning the Fabaceae family with the presence of two supplementary genera namely *Colophospermum mopane* and *Peltophorum africanum*. The vegetation that this study has been able to reconstruct for this site is mixed bushveld dominated by species from the Combretaceae and Capparaceae families. *Colophospermum mopane* is found in the *Colophospermum mopane* woodland/ *Colophospermum mopane* veld of the Mopane savanna vegetation. It is a tropical plant that falls within the Zambezian Domain of the Sudano-Zambezian Floristic region[9] (Eichhorn and Vogelsang 2011). Today the mopane formations cover large areas of the northern and north-eastern parts of South Africa, Botswana, Zimbabwe, Mozambique and northern Namibia with a total surface estimated to almost 555 000 km^2 (Smit and Rethman 1998). The vessels of *Colophospermum mopane* are arranged in radial multiples, the thin rays are 1–3 seriate. The charcoal may resemble that of *Vachellia/Senegalia* type 2 but *C. mopane* has more axial parenchyma than these taxa.

Peltophorum africanum (weeping wattle) falls within the Mountain bushveld, which is also the case for some *Combretum* and *Terminalia* species. The *P. africanum* archaeological charcoal shows solitary vessels and 1–2 seriate thin rays. *Peltophorum africanum* is traditionally used to treat a number of ailments including diarrhoea, dysentery, helminthosis and as an immunity booster for both animals and animals (Bizimenyera, 2006). Jacobson et al. (1995) portray the vegetation around the site today as being within the mopane woodland with elements also belonging to the thornveld and mountain savannah. Noticeably, except for mopane, the site's woody taxa are very similar to those of Leopard Cave, and likewise very similar to today's vegetation cover around the site. Today *Peltophorum africanum* is observable close to the site of Leopard Cave.

6.4. Arboreal species from seed analysis

Woody species of the mixed woodland/shrubland communities composed by *Boscia foetida* and *Grewia* have the highest representations among the identified woody taxa on the basis of seed/fruit remains. A thorny bushveld is also attested through the presence of *Commiphora* and *Vachellia/Senegalia* species. Urso et al. (2013) mention a *Ziziphus –Berchemia* association comprising other woody taxa such as *Colophospermum mopane* and the woody *Commiphora* species in northwestern Namibia. *Berchemia-Ziziphus* vegetation occurs in valley areas as well as along drainage lines. The main riparian species found here is *Euclea pseudebenus* that may have grown along the Ugab River.

6.5. Annual plants from seeds analysis

The presence of *Cyperus* seeds suggests once more the proximity of riverine plant communities. Sedges, however, do not necessarily need a constant mesic environment to thrive. Their presence at the contemporary sites of Big Elephant Shelter and some Middle Stone Age sites is for example not associated with perennial streams (Wadley 1979, Sievers et al. 2011). The sedges at these sites are believed to have been used as bedding and the grains and corms, for human nutrition. The presence of sedges is noted today in most parts of northwestern Namibia including at Big Elephant and they also exist in the dry streams near Leopard Cave. *C. fulgens* is one of the most common species in the northwestern parts of Southern Africa today.

Lapeirousia is another of the plants whose corms are consumed. It was noted in the archaeological record of Geduld but also grows around the site today and in the Erongo region (Wadley 1979).

The remaining taxa belong to the Fabaceae family (*E. elephantina* and *T. esculentum*) and correspond to herbaceous xeric taxa with foliage that dies off in winter and that resuscitates with the new rains. Together with *C. lanatus* these constitute very important legumes for arid-environment dwellers.

6.6. Toteng's arboreal species from charcoal

The low quantities of charcoal available from Toteng are responsible for the lower diversity of taxa obtained through the charcoal analysis. Still the results suggest that the site was surrounded by Mixed woodlands environments comprising taxa from the Capparaceae and Combretaceae families as well as Arid Sweet woodlands with thorny *Vachellia/Senegalia* species associated with *Dichrostachys cinerea* (Figure 64).

Different phytocenoses of the arid savannahs and other semi-arid regions are controlled by a number of factors, especially soil types and annual precipitation (Van Rooyen and Van Rooyen 1998, Palmer and Staden 2006). In the context of Southern Africa, this type of vegetation occurs in arid but fertile areas that are usually characterised by fine-leaved vegetation. However, the *Colophospermum mopane* around the study area, i.e northwestern Namibia

[9] The vast stretches of woodland, savannah and grassland vegetation with occasional dry forests and thickets, and patches of edaphically controlled swampy vegetation, in a wide zone in Sub-saharan Africa around the Guineo-Congolian Region (Weger 1978)

Figure 64. Vachellia bushveld near the site of Toteng. Photo: M. Mvimi.

and the central Kalahari around the Okavango Delta, is rather *broad-leaved.* In northern Namibia, the mountainous areas amidst savannah are complemented by *Commiphora* shrub savannah. Eichhorn and Vogelsang (2011) show that the distribution of mopane is restricted to the limits of the Kgalagadi basin where the sites of Geduld and Leopard Cave are found and also in the central Kgalagadi where the site of Toteng is located.

In conclusion, the macrobotanical remains from the three sites tell similar stories:

1. One of LSA communities who gathered vegetal resources from similar vegetation communities; vegetation with a composition comparable to what is found in these areas today. This hypothesis is supported by the presence of woody trees that conform to different forms of woodlands and a broader wet and dry open savannah. These woodland taxa include *Combretum apiculatum*, *Combretum molle*, *Terminalia prunioides* and possible *Boscia albitrunca* with a significant dominance of the Arid Sweet woodland that is characterised mostly by the genus *Vachellia/Senegalia* at the three sites. Though they are absent from Toteng, exceptional woody taxa, namely, *Colophospermum mopane* and *Peltophorum africanum*, are present in Geduld while *Sclerocrrya birrea* could be found only in Leopard Cave. These environmental similarities are also observable from woodlands comprising shrub species such as *Grewia* sp., *Boscia foetida*, *Commiphora* and *Vachellia/ Senegalia* species that are adapted to valleys, mountains and mountain slopes. Plants that belong to the stream bank communities such as those in the genera *Euclea* and *Cyperus* are found at Geduld. And finally, grasslands (Poaceae), xeric herbaceous species like *C. lanatus,* and leguminous perennials like *T. esculentum* and *E. elephantina* from both Geduld and Leopard Cave.
2. The macrobotanical results show very similar climatic conditions across the Kgalagadi Basin; results which are in harmony with the semi-arid/savannah Holocene climate of the savannah and Nama Karoo biomes discussed in chapters 1 and 2. Therefore indicating that climatic conditions have changed very little or in an insignificant way as far as the study period is concerned.

6.7. Making inferences from today's vegetation

The present-day environment of the two Namibian sites is relatively similar. Both are located in the proximity of escarpments and near seasonal streams; the Ugab river flows near Geduld and a small stream in the valley at the foot of the basalt massif of the Erongo is easily accessible from Leopard Cave. Big Elephant is close to both the Omaruru and Khan rivers to the north. Toteng, on the other hand, is an open site at the mouth of two rivers within the alluvio-lacustrine soils, and prone to flooding from the Okavango Delta. Moreover, the area receives slightly more rainfall than the other two sites. Toteng is found in a region with low open grass shrublands/woodlands and tall open shrublands/grasslands with riparian reeds and sedges as well as *Vachellia/Senegalia* thickets where the sandy substratum influences the vegetation cover (Vanderpost et al. 2015).

Whereas the species growing around the Namibian sites and Toteng are in many cases similar, the impression given by the vegetation at Toteng is that of much larger trees whose growth is probably favoured by better soils and a higher water table. Plant dimorphism was also observed by Leakey et al. (2005) by comparing the Namibian and South African marula for possible cultivation where they stated: "The fruits of 'Namibian Wonder' Marula were, however, very much heavier". (See chapters 2 and 4 for detailed descriptions of past and present vegetation in relation to the sites.)

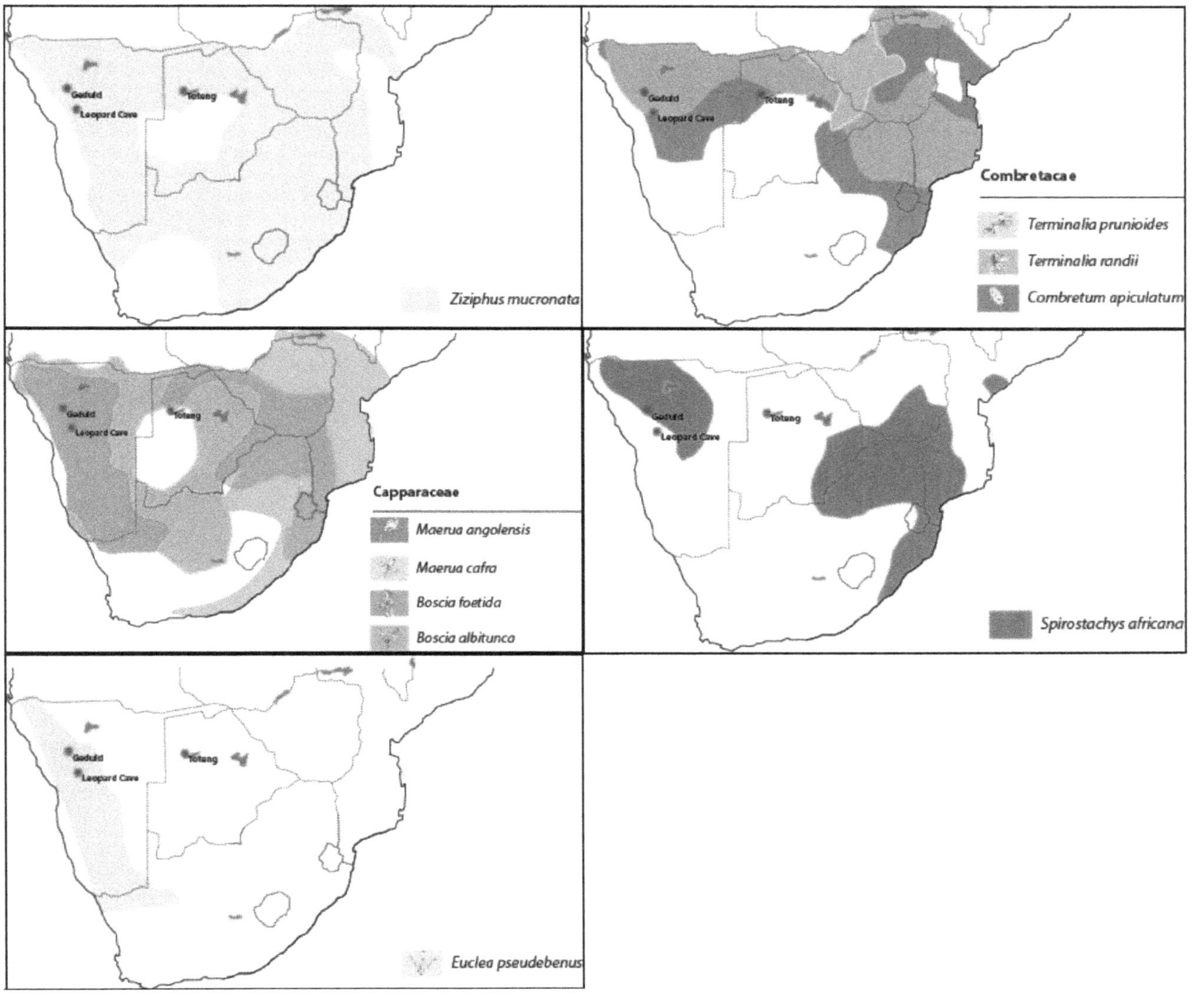

Figure 65. The current distribution of some of the major plant communities found on the sites. Photo: M. Mvimi.

6.8. Wild fauna to complement the environmental setting

The three study sites have delivered well-preserved and diverse faunal remains that in parallel with the plant remains inform us on past environments and subsistence strategies (Thackeray 1979).

At Leopard Cave, the mammal record is dominated by small bovines such as impala and klipspringers as well as species adapted to mountain environments such as the rock hyrax (*Procavia capensis*). The reptilian fauna comprises snakes and the monitor lizard. Guinea fowl and ostrich were hunted birds.

At Toteng large bovids like buffalo (*Syncerus caffer*) and red hartebeest (*Alcelaphus buselaphus*) have been identified. Among the smaller bovids appear impalas, grey duiker (*Sylvicapra grimmia*), and steenbok (*Raphicerus campestris*). Wet environment hosted sitatunga (*Tragelaphus spekei*), lechwe (*Kobus* cf. *lechwe*), hippo (*Hippopotamus amphibious*) and crocodile (*Crocodylus niloticus*). Tortoise, snakes and monitor lizard, as well as crab, bullfrog and gastropods, are also existent. Finally, baboon and the vervet monkey as well as carnivores such as lion complete the rich faunal assemblage at Toteng.

At Geduld, dominant bovids range from small-sized antelopes like the Damara dik-dik (*Madoqua kirkii*) to the common duiker (*Sylvicapra grimmia*), steenbok (*Raphicerus campestris*) and the larger hartebeest. (*Alcelaphus buselaphus*). Bones from domestic sheep and cattle are attested from ca. 1790 + 50 BP. Other types of fauna consist of birds (guinea-fowl, dove, francolin) and reptiles (*Varanus niloticus,* snakes…). While the bullfrog might have come from the river near the site, the presence of seal phalanges came as a surprise to the archaeozoologists as it must have been brought from the sea.

Other sites such as Big Elephant Shelter, have provided similar faunal assemblages (Wadley 1979). LSA layers at the site of Apollo 11 have yielded primate remains such as baboon, such bovids as steenbok, springbok and gemsbok as well as such carnivores as the jackal and hyena. Warthog, rabbit and the mountain zebra are equally present. Ostrich, snakes and some unidentified fish are found (Thackery 1979).

Indeed, the results from this faunal data from northwestern Namibia and central Kgalagadi are quite similar. Only small variations are perceptible according to the different environmental settings where the sites are found. In general, the wild fauna is composed mainly of grazers such as antelopes, hippo, equids and animals that are both grazers and browsers like the *Procavia*. The spectra from the mountainous Leopard Cave and Geduld areas encompass rocky areas where species like klipspringer and *Procavia* would have lived. Grazers are found at Toteng too but since it is in the proximity of wetlands, Toteng fauna includes hippo, lechwe, sitatunga, shellfish and fish. Together with the sites of Apollo and Big Elephant, the fauna observed here is concomitant with open plains, woody and grassy savannah environments with rocky outcrops.

Moreover, these animals are found around the area today (Figure 66). Ungulate communities in these areas include such large browsing fauna as the black rhinoceros, kudu and giraffe, which live in woodland vegetation. Springbok, zebra, gemsbok, impala and ostrich also occur on the open plains. The mountain zebra (*Equus zebra hartmannae*) occurs on the inland mountainous plateaux which run parallel to the South West African coastline, with klipspringer and dassie common on rocky outcrops. The more diverse ungulate faunal community in northern Namibia and north-western Botswana can also be ascribed to the fact that temperature variation in these areas is less extreme than in the southern regions of the country (see chapter 1 for full southern African fauna).

These open tree, grassland and shrubland environments, even though relatively arid, then and now, support a rich biological diversity. Therefore, it could be asserted that the LSA populations living in the study areas had a wide range of resources to choose from to sustain their subsistence needs.

6.9. Subsistence strategies and use of biological resources

The lack of systematically interpreted archaeobotanical data from various LSA sites creates an imbalance, because these LSA communities tend to be portrayed as predominately hunters. This study provides information not only on the environmental conditions but also on the human-vegetation relationship; A relationship from which certain patterns about vegetation choice is evident.

6.10. The choice of species

A very interesting dynamic emerges between charcoal and seed exploitation traits. The two are comparatively different with wood taxa exploited for combustion that differ from those trees bearing edible fruit. The only taxa that are found in both the seeds and charcoal samples are those from the Capparaceae family, and solely from the site of Geduld. Diagnostic charcoal taxa from these families are either from the genera *Boscia* or *Maerua* while the seeds from the Capparaceae families are exclusively from *Boscia albitrunca* and *Boscia foetida*. Capparaceae (*Boscia and Maerua*) charcoal also exists in Leopard Cave, and an unidentifiable type is found in Toteng.

Apart from underlining vegetation diversity, this indicates local and differential sourcing of vegetal resources with a distinction in the choice between fruit-bearing trees and shrubs and non-edible taxa. Could this reveal a strategy to safeguard fruit trees useful as a food resource by avoiding cutting of their wood? Unfortunately, it is difficult to tell from the charcoal record whether the wood was collected dry or if trees were felled for the purpose of combustion. Another possible factor that could influence the choice of species may have been taboos that prohibit the use of certain taxa considered sacred, poisonous or in general improper for certain uses. Similarities and differences in the spectra of chosen species also suggest they could be linked to cultural groups (as it is the case among present-day communities; see chapter 4)

6.11. Two exceptional species of Colophospermum mopane and Sclerocarya birrea

While the presence of *Colophospermum mopane* in Geduld is unsurprising, its low frequency in the archaeological record for a plant with the provenance of the Mopane woodland is unexpected. It is possible that it could have been "preserved" the same way fruit trees have been. *C. mopane* is a highly resourceful plant and much appreciated for its many uses. The strong wood is used for making

Figure 66. Vegetation of the Erongo showing Kudu (l) and zebra (c) in the thorny shrubland near Leopard Cave and an ostrich family in northwestern Botswana. Photos: M. Mvimi.

poles, stools, wood carvings and kitchen utensils (Van Wyk and Gericke 2017). The firewood is long-lasting, the gum can be melted and used as glue, and the cambium weaved into rope or used as tannin. The inner cambium is also known to have medicinal properties. The tree hosts edible caterpillars (*Gonimbrasia belina*) commonly referred to as mopane worms. Mopane is not used just because it is the dominant species, but because it is a highly preferred resource with many uses (Kennedy and Potgieter, 2003; Coates Palgrave, 2002).

Sclerocarya birrea (marula) appears at Leopard Cave but in very small quantities. This is a tree that thrives in different woodland conditions, on variable types of soils, especially when well drained and with an annual rainfall ranging from 200–1500 mm (Cheikhyoussef and Embashu 2013). The period around 2600±150–2100±100 BP corresponds to these environments, raising the possibility that its existing in this mixed woodland cannot be excluded. But as with mopane, hard to explain is its occurence in very small amounts and the fact that it is non-existent in the previous phase (3700±150–3200±150 BP) despite similar settings between the two phases. Like mopane, *S. birrea* is a plant of many uses, even earning itself the term 'power' plant (Eloff 2001). The fruit of *S. birrea* is a part of many rural diets but also of great economic value. The fruit is used to make jam, traditional beer, and can be eaten fresh (Emanuel et al. 2005). The leaves are eaten as relish by the Pedi of South Africa, the kernels are eaten raw and oil is extracted from them. Its branches are used as firewood and the bark has antibacterial properties (used to treat fevers and a wide variety of stomach ailments including ulcers) and has over the years been relied on by a variety of southern African traditional healers. The bark also contains tannin and tanning material (Eloff 2001). Namibian fruits were significantly larger than those from South Africa (Cheikhyoussef and Embashu, 2013).

The single occurrence of marula charcoal together with the absence of its fruits may suggest that the plant did not grow in the area during the LSA period. In the reconstruction of the vegetation cover in the adjacent Kunene region of north-western Namibia, conducted by Eichhorn & Vogelsang (2011), this species is also absent from the charcoal record. It is thus possible that the plant was brought to the site from elsewhere (Walker, 1995).

Long-distance transport of certain commodities is further attested by the presence of several seal phalanges in the faunal assemblage at Geduld, located at a distance of 270 km from the nearest coast.

Two other plants, namely *Tylosema esculentum* and *Citrullus lanatus,* are not rare in the archaeological record but they are worth noting as they have played a major role in the survival of arid environment dwellers over the years including today. The large tubers of *Tylosema esculentum* are roasted and eaten and can be eaten raw as a thirst quencher (Van Wyk and Gericke 2017), for example by the Himba people. The beans which contain around 30 percent protein, are eaten together with the leaves. The San eat the leaves raw and roast the seeds and roots, and the Tswana of southeast Botswana consume the seeds (Sullivan, 1998). *T. esculentum* seeds and leaves can also be used as livestock feed.

Tsamma melon contains less than 7 percent sugar and has a low concentration of minerals and vitamins (B and C). The protein-rich seeds contain edible oil and are commonly roasted. Tsamma melons are eaten as fresh fruit or can be boiled or steamed as a vegetable. In mid to late summer in dry areas Tsamma is valued as a source of water (Mujaji, 2009). The tender young leaves of *Citrullus lanatus*, as well as the fruits, are cooked like spinach or green vegetables, while the fruit flesh may be cooked as porridge accompanied by a starchy side.

Human remains are rare on Late Stone Age sites in Southern Africa and it is thus difficult to draw conclusions concerning the population structure and various aspects of health, diet and daily life that can usually be read from bone remains. It is generally accepted, though, that these groups were small and mobile(Palmer & Staden, 2006). The role of domestic livestock during the LSA has been much debated (see chapter 2) and the rarity of faunal remains formally identified as belonging to cattle or caprines has been explained in various ways. According to the study of charcoal and seeds, the vegetation communities around the sites did not change much through time. Moreover, they seem to have been very similar to present residents in their use of botanical sources. Thus, it is possible that the presence of LSA communities, even in a rather fragile environment, did not have a large impact on the local plant communities. It is not until recently that human-induced changes have altered vegetation patterns and composition mostly due to farming and herding practices (Carrión 2000). Even in Toteng and Geduld where domesticated animals have been present for some time, the impact of grazing is not visible in the anthracological record. This might imply that livestock was of minor importance or simply not really "ordered" in these areas, an assumption that would also explain the small amount of bone remains from domestic fauna at the sites and the persistence of hunting a large array of wild animals.

6.12. Seasonality of fruit collection and possible processing

Fruit was necessarily collected when available (Table 25), and seasonal changes likely influenced the cave settlement. The plant remains offer information on pharmacopoeia and seasonal food availability (See Ethnography chapter for more detail). Jacobson et al. (1995) schematised the fruiting seasons of the plant spectra at Geduld in order to understand their exploitation throughout the year (See chapter 2 for more detail). Here a similar model was adapted for the general seed/fruit corpus of this study to discern the relationship between the fruiting season, fruit harvesting and site occupation. This is a summer rainfall area and most of the fruit ripens between the summer

Table 25. Seasonality of fruit trees reflecting the macrobotanical spectra

Season		**Summer**			**Autum**			**Winter**			**Spring**			
Fruiting period (month)		Dec	Jan	Feb	Mar	Apr	May	Jun	Jul	Aug	Sep	Oct	Nov	Dec
Anacardiaceae	*Sclerocarya birrea*			*	*	*	*	*						
Burseraceae	*Commiphora*		*	*	*	*	*							
Capparaceae	*Boscia albitrunca*											*	*	*
	Boscia foetida			*	*	*								
Combretaceae	Terminalia sericea			*	*	*	*	*	*	*				
	Terminalia prunioides			*	*	*	*	*	*					
Ebanaceae	*Euclea pseudebenus*			*	*	*	*	*						
Olacaceae	*Ximenia americana*	*	*	*										
	Ximenia caffra	*	*											
Rhamnaceae	*Berchemia discolor*		*	*	*	*	*	*	*					
	Ziziphus mucronata				*	*	*	*	*	*	*			
Rubiaceae	*Vangueria infausta*		*	*	*	*								
Tiliaceae	*Grewia flava*		*	*	*	*								
	Grewia flavescens		*	*	*	*								
Curcubitaceae	*Citrullus lanatus(ecirrhosus)*			*	*	*	*							
Cyperaceae	*Cyperus* sp.													
Euphorbiaceae	*Ricinus communis*						*	*	*					
Fabaceae	*Elephantorrhiza elephantina*	*	*	*	*									
	Tylosema esculentum			*	*	*	*	*						
E. elephantina and *T.esculentum* vines die and the perennial tube remains dormant for the rest of the year and new sprouts appear with the rain														

and autumn except for *Boscia albitrunca*, which bears fruit in the spring. Harvesting fresh fruit and possibly processing fruit is present in the evidence. Results here support Jacobson et al. (1995), that noted phases of seasonal occupation at the site, probably during autumn until winter. There is also evidence that seed processing, such as *Ximenia* and melons (*Citrullus* spp.) was also done at Geduld and Leopard Cave. Systematically broken endocarps, as well as whole fruit with dry skin, are present. Preparation requires drying of fresh fruit and breaking of the endocarp to access the kernels containing oil. Therefore the communities in this area may have used the caves as food processing stations or just processed the fruit during their period of stay (Urso 2013). Comparison of the two sites as being used for similar purposes is impossible as Leopard Cave has evidence of continuous occupation over years, while further excavation needs to be done at Geduld to establish a continuous sequence.

The *Ximenia* ripe fruit (Figure 67, Figure 68) has a vitamin C content of 27%, is high in potassium and contains protein. The seed has a 65% oil content. Fruits have a refreshing sour taste, best eaten when over-ripe, but can also be used for making jam, desserts and jelly, an alcoholic beverage, or can be added to porridge (Van Wyk and Gericke 2017). Oil from the seed is used on human skin, lamps and for working animal hides. The nuts can also be eaten. Oil from the seed is sometimes used by Himba and Tjimba of Kaokoveld as a substitute for animal fat which is normally applied to the body (Sullivan, 1998).

Even though chances of finding preserved botanical material from archaeological sites are not always assured, the taphonomical nature of certain sites create exceptional preservation conditions. Firstly, the chances of carbonised wood to survive in an archaeological site are very high, and the inert nature of the charcoal allows it to survive (Esterhuysen & Mitchell, 1996). Seeds/grains alike can be preserved by carbonisation, by desiccation or mineralisation.

The exceptional preserved nature of the macrobotanical remains at the sites of Leopard Cave, Geduld and Toteng have allowed for their identification. The analysis from these charcoal and seeds/grains have furthered the understanding of socio-environmental and possibly sociocultural behaviours pertaining to the relationship between LSA communities and their environment circa 2000 BP. While the study sheds light on paleoecological and paleobotanical changes, it raises a number of questions concerning both the environment and subsistence practices.

This book explores the nature of the environment during the period when the first herders arrived in the southern

Figure 67. An example of *Ximenia* for oil extraction in Angola and a fresh *Ximenia* fruit. (Urso et al. 2006, M. Mvimi.)

Figure 68. Split *Ximenia sp.* endocarps. Photo: M. Mvimi.

African region around 2000 years ago. In particular it sought to determine how these communities exploited the environment and how the environment influenced their food procuring habits.

The macrobotanical assemblage from Leopard Cave and Geduld show a well-defined variety of charcoal and seed taxa, and a similar intra-site spectrum of the charcoal. Toteng, though less diversified, shows taxa similar to that of Leopard Cave and Geduld. Based on the available data, the following information could be extrapolated; the peripheries of the Kgalagadi basin in northwestern Namibia was comprised of woodlands, shrubs and open grasslands dominated mainly by *Vachellia/Senegalia*, and species of the Combretaceae and Capparaceae families. The Central Kgalagadi appears to have had similar environmental conditions because the Toteng area also comprises woody *Vachellia/Senegalia* taxa, and taxa of Capparaceae and Combretum families.

The communities living in these hinterlands procured food plant resources from their immediate environs. Fruit and wood may have been collected in the local area but with a clear, selective approach. A comparison between the charcoal and seeds assemblages from Leopard Cave and Geduld show that those species used for combustion were not the same species that bore edible fruit, except for a few *Boscia* species. The possible introduction of some species like the Marula in the phase between 3700±150 – 3200±150 cal BP is likewise plausable.

Archaeobotany extends beyond plant identification by correspondingly considering people as the actors in their own environments. Working in Namibia where very few detailed paleobotanical research has been done and in Botswana, were there is none, seeking relevant sources to infer is indispensable. First, an ethnographic approach was included for the main reason that oftentimes it is very difficult for archaeobotanists to interpret plant use in the archaeological contexts, especially for earlier periods in places where people gathered resources rather than cultivated and stored them.

Ethnography is additionally important to make deductions especially for the vegetal connotations that are associated with intangible aspects, such as taboos. Interviews among the San and Damara communities in the Erongo aided with such. The San communities are more centred around "rules" concerning general plant use and are more selective than the Damara. The Damara, on the other hand, exploit almost all the plants in their environment, and they mentioned fewer restrictions. To address the challenges associated with doing research in areas where previous archaeobotanical investigations are absent; this study included the construction of a detailed basic reference collection, mainly to facilitate identification. The processed wood and seeds were collected around the study sites and a catalogue was compiled. Archaeobotany and ethnobotany alone cannot give a full picture of the LSA in Southern Africa, but this study has managed to show that the late Holocene environmental conditions in the Erongo were similar around 3500 years ago to what they are today and that the local communities exploited their environment and selected between their fruit and combustion material. The similarity of the environment from then and now is supported by the zooarchaeological record from the same sites (Pleurdeau et al., 2012; Jacobson, 1987; Jacobson et al., 1995) and those from the environs (Thackeray, 1979; Wadley, 1979). The archaeological fauna found in these sites includes mainly small and large grazers. The majority if not all can be found in the area today. This faunal information together with the botanical information shows that the area was very rich in food and other biological resources that these people exploited. Some of the plants and animals may have been brought to the site, as was the case with *Sclerocarya birrea* in Leopard Cave and the seal bone in Geduld. For a period spanning the last 2000 years where the environments in Southern Africa are anticipated to be marked by cultural events that might have had an impact on the environment, such is not the case in the study area concerned here. Questions arise thus about this lack of impact by "herding" on the environment. Is it because people kept very small stock? Finally, the lack of representation of starchy foods or grasses in the assemblages creates an imbalance.

The Archareobotanical approach employed in this study has substantiated the ability to make a significant input into our understanding of LSA people and their environmental contexts at the period. Extending such work will be indispensable to other LSA studies. The study corpus needs to be expanded at sites in addition to Geduld and Toteng, where the assemblages are not very large. Because resemblances between the study sites and other sites in the region have been established, archareobotanical research needs to be extended to include additional sites for comparison, to determine relationships, and to gain more insights into Southern African LSA communities and their vegetal surroundings. Dietary evidence can be expected to add to this understanding. Fruit and wood exploitation practices were attested, but noteworthy is the absence of the evidence of the use of grasses (Poaceae) at the three sites. While the preservation conditions of grasses are not recognised during this study, LSA dietary habits that included carbohydrate acquisition and intake cannot be disqualified. Therefore, specific methodological approaches such as starch analysis are essential. As is the case with the broader archaeobotanical prospectives, the inclusion of neighbouring sites can be useful to question whether grasses do indeed occur. Finally, as part of this study, a botanical reference collection was built, and a preliminary catalogue compiled that proved very useful for identification reasons. This collection can serve as a foundation upon which to create a more comprehensive reference database for the Southern African arid and semi-arid milieus.

Appendix 1

Wood Anatomy for Identification Purposes

The identification process in this chapter is exclusively based on hardwoods/angiosperms, the only type available in the charcoal assemblage. In botanical terms, angiosperms are vascular, flowering plants that make over 80 percent of all living plants. Angiosperms comprise a large variety, among others shrubs, aquatic plants, bulbs, herbs, epiphytes and trees (Simonin & Roddy, 2017; Christenhusz & Byng, 2016). A living, growing tree has two main domains, the outer shoot and the underground roots. One of the components of the shoot is the stem, which comprises a variety of layers (Figure A1) the outer bark, which is the protective layer of the tree also helps reduce the loss of water by evaporation, the phloem, moves sucrose sugars and proteins, from the leaves where they are produced by photosynthesis to the roots and the rest of the plant. The vascular cambium which is situated between the wood and the bark is where cell division occurs, the cells either become wood or bark cells. The sapwood is responsible for the storage and synthesis of biochemicals as well as the conduction of sap. At a given moment, the need for the conduction of sap to the leaves by some parts of the trunk ceases. At this stage, the vessels in the centre of the plant fill with extractives and minerals resulting in the heartwood and the pith.

For all wood identification criteria, identification of charcoal from archaeological contexts requires careful attention. This is due to the fact that even though the inert structure of the charcoal is mainly conserved after carbonization, fire and other post-depositional factors including pest-activity, may possibly interfere with the wood structure, and different species are subject to react to these factors in different ways. Therefore, it is possible to convey information on collection strategies, preferred parts of the wood and wood defence mechanisms (e.g. reaction to insect attack) as well as different taxonomic information. Furthermore, identification is based on xylem which contains the characteristics described below. (Prychid et al.,2004; Dabney et al., 2016). Notably, there are several other characteristics applicable to different other species but may not be conjured here as they are not pertinent to the corpus at hand.

Below are the wood anatomical features and their functions according to the International Association for Wood Anatomists:

- **Rays**: Transport water and other dissolved nutrients across the stem.

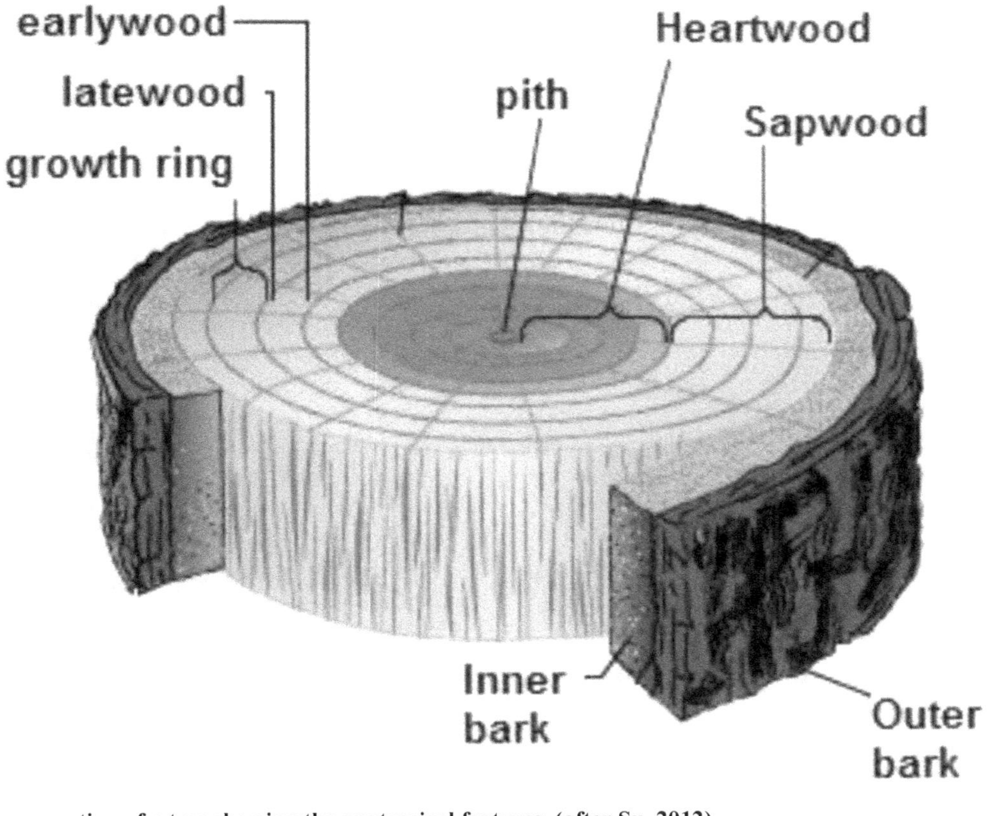

Figure A1. Cross-section of a tree showing the anatomical features. (after Su, 2012).

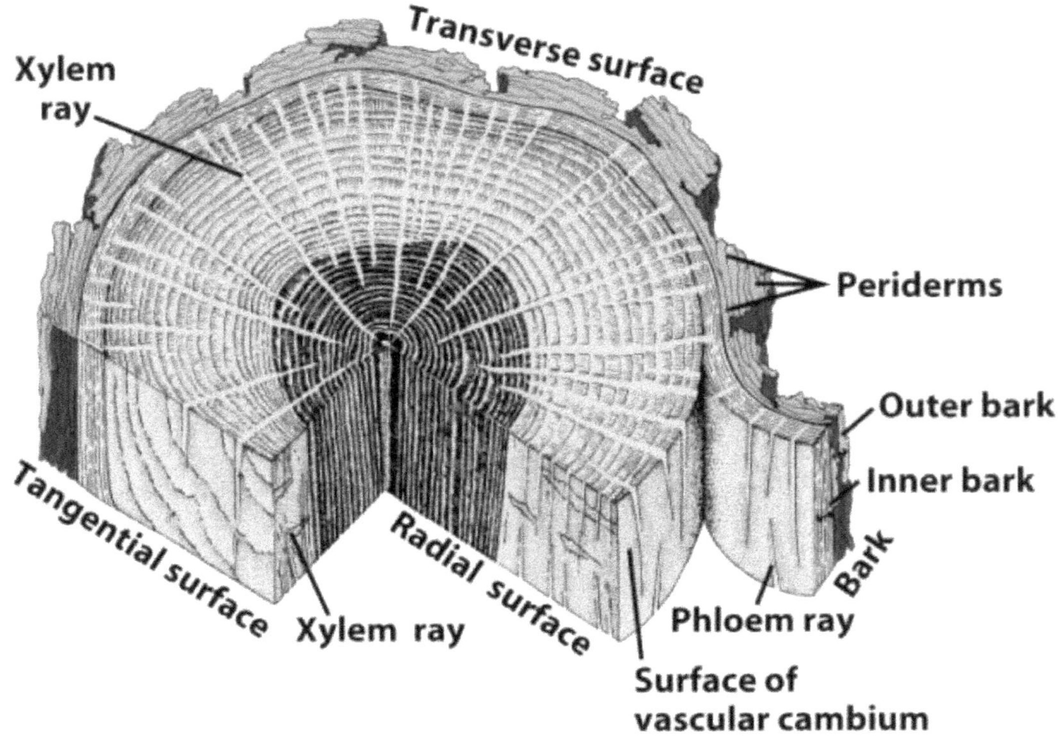

Figure A2. Wood phases and wood charcoal showing the below discussed features. Top:(after Raven, et. al. 2005; © W.H. Freeman and Company); bottom: M. Mvimi.

- **Parenchyma**: Thin-walled cells responsible for the storage and distribution of food. In hardwood, parenchyma cells are brick-shaped and found around gum canals and ray cells.
- **Fibres**: Thick-walled cells that mainly serve to support and strengthen the trunk.
- **Tracheids**: they serve both the mechanical and conductive needs of softwoods. They make 90% of the volume of the wood.
- **Xylem** consists of narrow tracheids and wider vessel members that form vessels.
- **Tyloses**: In some instances, vasicentric balloons into the interior of adjacent vessel elements. The tyloses efficiently block unused vessel as passageways for incursions like insects and fungi.
- **Perforation plates**: Found at the ends of the vessels and they allow for the efficient transport of fluids. Vessels are formed by vessel elements piling one on top of the other, and the perforation plates are the areas where the vessel elements overlap. They could be reticulate, scalariform, simple etc.
- **Vessel**: made up of a row of meristematic cells of procambium or vascular cambium which remain attached end on end in a longitudinal series.
- The vessel patterns (presentation, porosity, groupings and arrangements) are used to distinguish the wood for identification purposes.
- **Intervessel pits**: The connections between adjacent vessel members.
- They could appear opposite, alternate or scalariform. Their shape also matters as a distinctive feature. Their size(μm) can also be used as a determining factor.
- **Vessel-ray** pitting: the connection between vessel elements and adjacent ray parenchyma cells. The size and border shapes are considered.
- **Helical thickenings/Spiral thickenings**: Helical thickening in the secondary xylem appears as a ridge of cell-wall material, usually deposited on the lumen side of the xylem in the tracheids of certain conifers, and in angiosperm fibres and vessels.
- **Phloem**: Inner bark. The function is to distribute manufactured foodstuff.
- **The tangential diameter of vessel lumina**: On the transversal section, vessels appear as open pores. Their diameters may vary from small (<30 μm) to large (>300 μm) but normally range between 50 and 200 μm.
- **Fibre wall thickness**: The fibre lumina size in comparison with the surrounding double wall thickness. The thickness of the fibre cell wall is the main determinant of density and strength. Plant species with thin-walled fibres have a low density and strength, while species with thick-walled fibres have a high density and strength (Wiedenhoeft and Miller 2005).
- **Sheath cells**: Cells found on the limits of broad rays. They are usually larger than the central ray cells.
- **Radial canals**: Pockets filled with resin. They could be thick or thin walled. The walls are not those of the canals themselves they are those of the cells

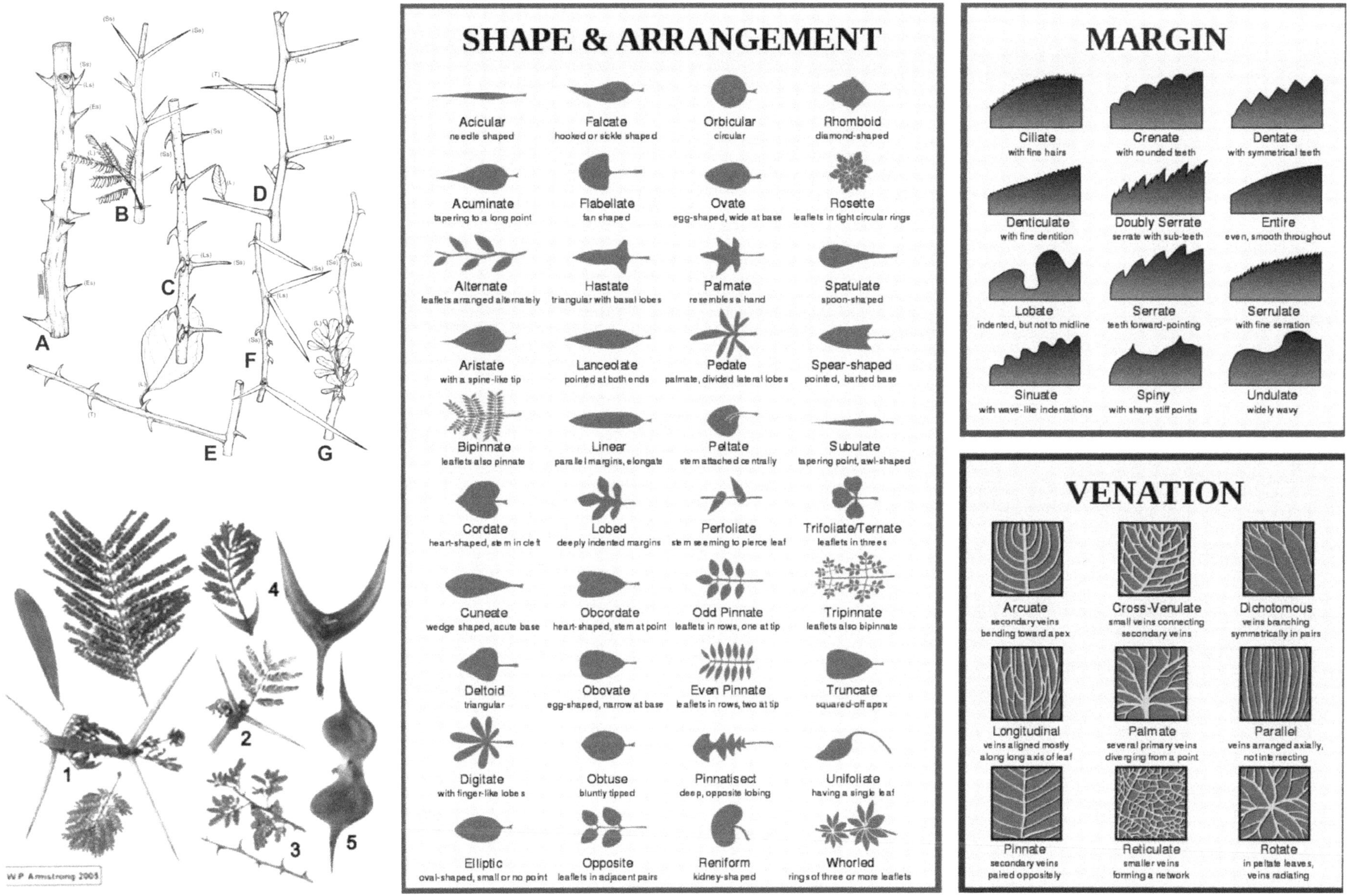

Figure A3. A description of the thorn and leaf morphology for tree identification and differentiation. (A) Prickles: (2 and B) Straight stipular spines: (C) Straight stipular spines and stipular hooks: (1 and D) Straight thorns (E and 3) Hook thorns: (F) Straight stipular spines and stipular hooks:(4 and G) Stipular hooks: Images adapted from; Right: sciencedendrologyproject.weebly.com, top-left: Charles-Dominique et al. 2016, figure S1, bottom left; Armstrong, 2005 and (https://www2.palomar.edu/users/warmstrong/plaug99.htm).

surrounding them. The canals themselves are not cells but cavities.

- **Prismatic crystals**: Mineral remnants of plant metabolism and chemical processes containing calcium oxalate.
- **Druses**: A form of crystals that have a star shape. They could be found in parenchyma and ray cells.
- **Silica bodies:** The plant takes in soluble monosilicic acid from the soil. The Silicon build up in the silica cells become the mineral structures of amorphous hydrated silica with various shapes and properties. They may be found in any part of the plant but they are mainly located in the ray or axial parenchyma cells.

Appendix 2

List of Identified Families Featuring in the Wood Atlas (According to Coates and Palgrave 2002; van Wyk and van Wyk 2007)

- **Anacardiaceae**

They are also known as the mango family. They are mostly trees or shrubs with compound leaves. Theoretically, they are expected to contain resinous latex. They have separate female and male species. Species include evergreen and plants like *Loxostylis alata* (Tearwood) or the *Harpephyllum caffrum* (Wild-plum) or deciduous species like the *Lannea discolor* (Live-long tree) and the *Sclerocarya birrea* (Marula).

- **Burseraceae**

Also known as the myrrh family, it is the family only represented by the corkwoods or *Commiphora* species with diverse leaf and bark morphology. The corkwoods are shrubs and trees characterized by their distinctive papery peeling bark and milky or watery sap some of which is aromatic. They are deciduous plants that range in size. Their fruit are spherical fleshy drupes with a hard, sometimes colourful stone that falls away when ripe. Their leaves are simple, compound uni-trifoliate, spirally arranged, grouped at the end of the branchlets and they produce a pungent smell when they are pressed. Species include the Purplestem corkwood *Commiphora multijuga*), Poison-grub corkwood (*Commiphora Africana*), Greenstem corkwood (*Commiphora neglecta*), Glossyleaf corkwood (Commiphora Schimperi), Paperbark corkwood (*Commiphora marlothii*) etc.

- **Capparaceae**

The plants in the caper family all contain mustard oil. Their flowers have four free sepals or petals and numerous stamens with elongated filaments. In Southern Africa, the Capparacae grow in areas of evergreen forest, open woodland, bush, scrub or sand forest. In this family, the ovary is borne on the gynophores, thus the fruit is marked by two sections separated by a scar left both on the receptacle and the perianth. The leaves in the Boscia species are in fascicles (tight clusters) along the branch and they are composed of digitately compound leaves with 3-5 leaflets. The Southern African genera found in this group include *Boscia* (shepherd's tree) *Maerua* (Spider bushes and bead bean trees) *Thilachium* (Cucumber-bush), *Cadaba* (Wormbushes), *Capparis* (Caperbush), *Ritchiea* (Ritcheas), *Bachmannia* (Fingerbush) and *Cladostemom* (Fingerbush).

- **Combretaceae**

The family is made up of the genera *Terminalia* (Clusterleaf) and *Combretum (*Bushwillow), both with winged fruit. The former has leaves in a spiral pattern, often in clusters and the fruit is surrounded by a single wing. They also have branches that deviate from the main shoot. The latter is characterized by opposite leaves and the fruit is usually four-winged. Combretums are the largest members of this family in numbers and they are key components in the bushveld vegetation of Southern Africa. And they are able to survive in poor sandy soils. The genera in this group include the Purple pod Terminalia (*Terminalia prunioides*), the Silver Terminalia (*Terminalia sericea*), Large fruit bushwillow (*Combretum zeyheri*), Combretum apiculatum, *Combretum imberbe*, *Combretum molle*.

- **Cucurbitaceae**

The gourd or the cucurbits family are mostly climbing or trailing plants with fistular, sometimes woody stems and they have tendrils that aid them in climbing. They have simple alternate, undivided leaves. These annual or perennial herbaceous plants consist around 975 species and 95 genera. Their root system is sometimes characterised by large tuberous roots. They have unisexual flowers. Seeds are often numerous, often compressed, smooth or variously pitted. Most of the Cucurbitaceae are cultivated and commercial plants but many wild species also occur. The wild species include the genera *Cucumis* and *Citrullus* and the most common is the *Citrullus ecirrhosus* (Tsamma/ Kgalagadi melon).

- **Ebanaceae**

Also known as the ebony family, the *Euclea* (Guarri trees) and *Diospyros* (Jackalberries) species in this family usually have opposite leaves that are whorled or sometimes spirally arranged with no stipules. The fruit are tough-skinned berries with a seed that is divided into 3 parts divided by a curved grove in Euclea and ellipsoid shaped with 2 lines joining at the apex in *Diospyros*. Present species include *Euclea pseudebenus* (Ebony tree), *Euclea undulata (*Common guarri) etc. *Diospyros lycioides (*Star-apple), *Diospyros acocksii* (Namaqua Jackal-berry), etc.

- **Euphorbiaceae**

The euphorbia family's main characteristic is the production of milky to watery latex. The' other distinguishing characteristic is the 3 lobed fruit that is crowned by 3 persistent stigmas that can either be dehiscent or indehiscent and fleshy. The genera include among others, different types of *Euphorbia, Spirostachys* and *Croton* species that are found around the study sites.

- **Fabaceae/Caesalpinaceae**

This group is also a legume of pod-bearing nature. They are recognizable by their butterfly-like leaves. They include the bauhinia group where we find, the Red bauhinia (*Bauhinia galpinii*), Yellow bauhinia (*Bauhinia tomentosa*), Butterfly leaf (*Adenolobus garipensis*) the Camelsfoot (*Piliostigma thonningii*) etc., the Mopane group that is characterized by the Mopane (*Colophospermum mopane*), Large copal wood (*Guibourtia coleosperma*).

- **Fabaceae/Mimosoideae**

Tree species in this group range from annual and perennial legumes to trees more than 16 metres high. They consist of different groups that are concomitant with the foliage of individual plants. The groups include, among others the "hook thorn group" with slightly curved spines and twice divided leaves. This group includes the Common hook thorn (*Senegalia caffra*), Umbrella thorn (*Vachelia tortilis*), the Flamethorn (*Senegalia ataxacantha*) or the Blackthorn (*Senegalia mellifera*). The "sweet thorn group" on the other hand consist of trees with straight thorns mostly occurring in pairs at each node but can occurring singly. The thorns are usually white and are placed at the base of the leaf. They include the Ana tree (Faidherbia albida), the Sweet thorn (*Vachellia karoo*), the Sicklebush (*Dicrostachys cinerea*) and the Camelthorn (*Vachelia erioloba*). The third group is the "False thorn group" which like the previous groups have twice-divided leaves but lacking thorns. Some of the species in this group possess large leaflets which are not so obviously bipinnate. Those in the Mimosoideae group include the Flatcrown (*Albizia adianthifolia*) and the Paperbark albizia (*Albizia tanganyicensis*). Those in the Caesalpinoideae group include the Wild Syringa (*Burkea africana*) and the African wattle (*Peltophorum africanum*). The tree species occur vastly in the savanna and arid northwestern, central and eastern southern Africa.

- **Olacaceae**

The sourplum family presents narrow-winged leaves of the *Olax* and the thorny-branched *Ximenia* groups. They are trees and shrubs that grow from 3 to 7 m high. Both genera have simple leaves arranged in the Ximenia group but they are alternate in *Olax*. The fruit in this group is a drupe ranging from 8 mm-2.5 cm depending on the *Olax* species and yellowish/reddish, ellipsoid and measuring 2.5 cm for the *Ximenia* species.

- **Poaceae**

Also called the Gramineae, this is the grass or the bamboo family that fall into the class Liliopsida (monocotyledons). These flowering plants' have stems that are hollow except at the nodes, where the leaves attach, and long narrow alternate leaves. The leaves are characterised by parallel veins and comprise a cylindrical sheath that wraps around the stem with the leaf-blade situated above it. The flowers are generally small, and the sepals or petals are not so evident. Of all the worldwide 10 000 species, about 1000 of them have been identified in Southern Africa with 300 of them being endemic. The rest have been introduced. A large number of grasses are found in Southern Africa. Even though the majority is found in grasslands, the Poaceae also occur in wetlands, forests and tundra environments. Domesticated Poaceae cereals found in Southern Africa include maize, sorghum, millet and sweet reeds and sugarcane. Wild grass uses include building material, food, as well as animal feed.

- **Rhamnaceae**

The Buffalo thorn or jujube group consists of shrubs and trees with glossy leaves and grow up to more than 20m high. The genera in this family include the thorny *Ziziphus* species eg Z. *mucronata* (buffalo thorn) and different *Berchemia* species *eg B. bicolor (*Bird-plum). The fruit are a drupe.

- **Rubiaceae**

The coffee and gardenia family are the largest family of trees in Southern Africa, with more than 10 000 species worldwide. The trees present leaves that are simple and sometimes opposite with entire margins. The leaves are rarely coiled or crisped. The stipules cover the growing tip and are usually interpetiolar. The genera include among others the most recognisable members of this group, i.e the wild medlar group (*Vangueria*) and the genera *Gardenia*.

- **Tiliaceae**

This is the jute and linden family of shrubs and trees that include several species from the "raisin bush group" notably *Grewia*. The leaves have several veins usually starting from the base, usually hairy and the edible fruit are drupes ranging from thin-fleshed to woody with longitudinal ridges. Species found in the region include Carpodiptera (*Carpodiptera africana*), Glyphaea (*Glyphaea tormentosa*), the Cape stock-rose (*Sparrmannia africana*) and the raisin bushes (*Grewia flava, G. flavescens*, *G. bicolor* etc.).

A4 - FABACEAE - *Senegalia mellifera subsp. detines*, Swarthaak (Erongo, Namibia)

IAWA Feature Code: 2,4,22,24,41,60,65,79,85,79,85,97,104,112,131,134,136,144,306

- Growth rings absent
- Wood semi-ring porous
- Simple plate perforations
- Alternate intervessel pits
- Intervessel pit sizes: 4µm
- Simple vessel-ray pits
- Mean tangential diameter vessel lumina: 83µm
- Vascular tracheids
- Paratracheal vasicentric parenchyma
- Banded/Confluent parenchyma (aliform in some areas)
- 1-2 seriate wide
- All rays procumbent(e)
- Slightly simply perforated ray cells
- Intercellular canal (mostly present in the cross section)
- Possible included phloem
- Resin canals also present
- Prismatic crystals in ray, fibre and parenchyma cells
- Druses present
- Crystal diameter: 14µm
- Crystal height: 20µm
- Crystals are present in long horizontal chains associated with parenchyma in the transversal
- Crystals also in radial multiples in the tangential side
- Septate fibres present
- Lowest ray cells at 4 cells high
- Highest cells at 14 cells high
- Included phloem

***Senegalia mellifera* tree. Photo: M. Mvimi**

a and d: transversal, b and e: radial, c and f: tangential

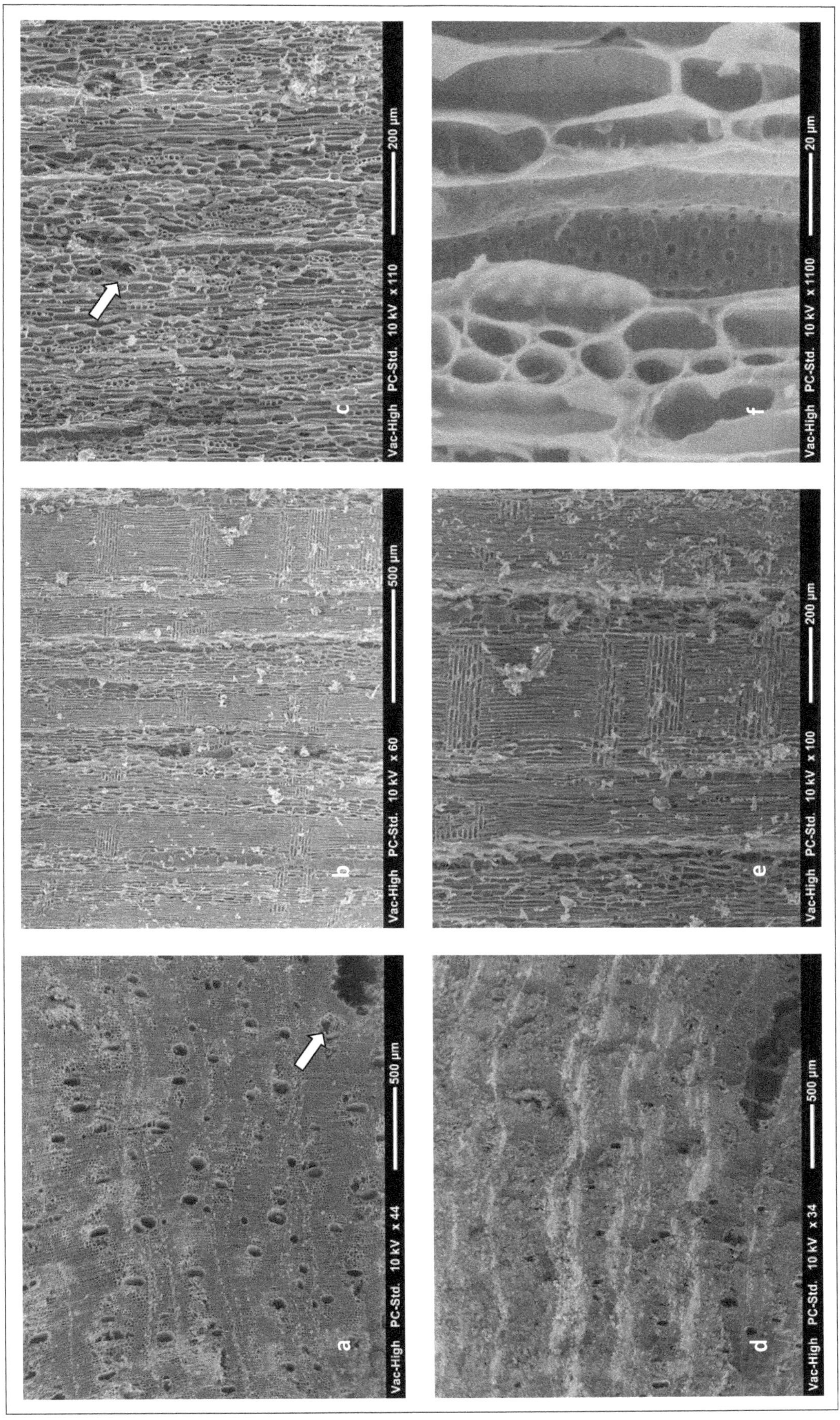

A5 - FABACEAE - *Vachelia robusta* (Burch) (Southeastern Botswana)

IAWA Feature Code: 2,4,9,24,13,30,36,40,66,69,83,97,130,106

- Wood semi-ring porous
- Vessels solitary, in pairs or in aggregates of 3
- Simple plate perforations
- Alternate intervessel pits (f)
- Vessel-ray pits similar to intervessel pitting
- Helical thickenings present
- Mean tangential diameter vessel lumina: 50µm
- Intervessel pit size = 2µm
- Non – septate fibres present
- Medium fibre walls
- Paratracheal confluent axial parenchyma
- 1-3 seriate
- Rays with scalariform pattern in tangential section
- Rays up to 60 cells high
- Shorter rays 6 cells high
- Resin canals present (round shape appearance) (e)
- Body ray cell procumbent with one row of upright cells

***Vachelia robusta*. Photo: M. Mvimi**

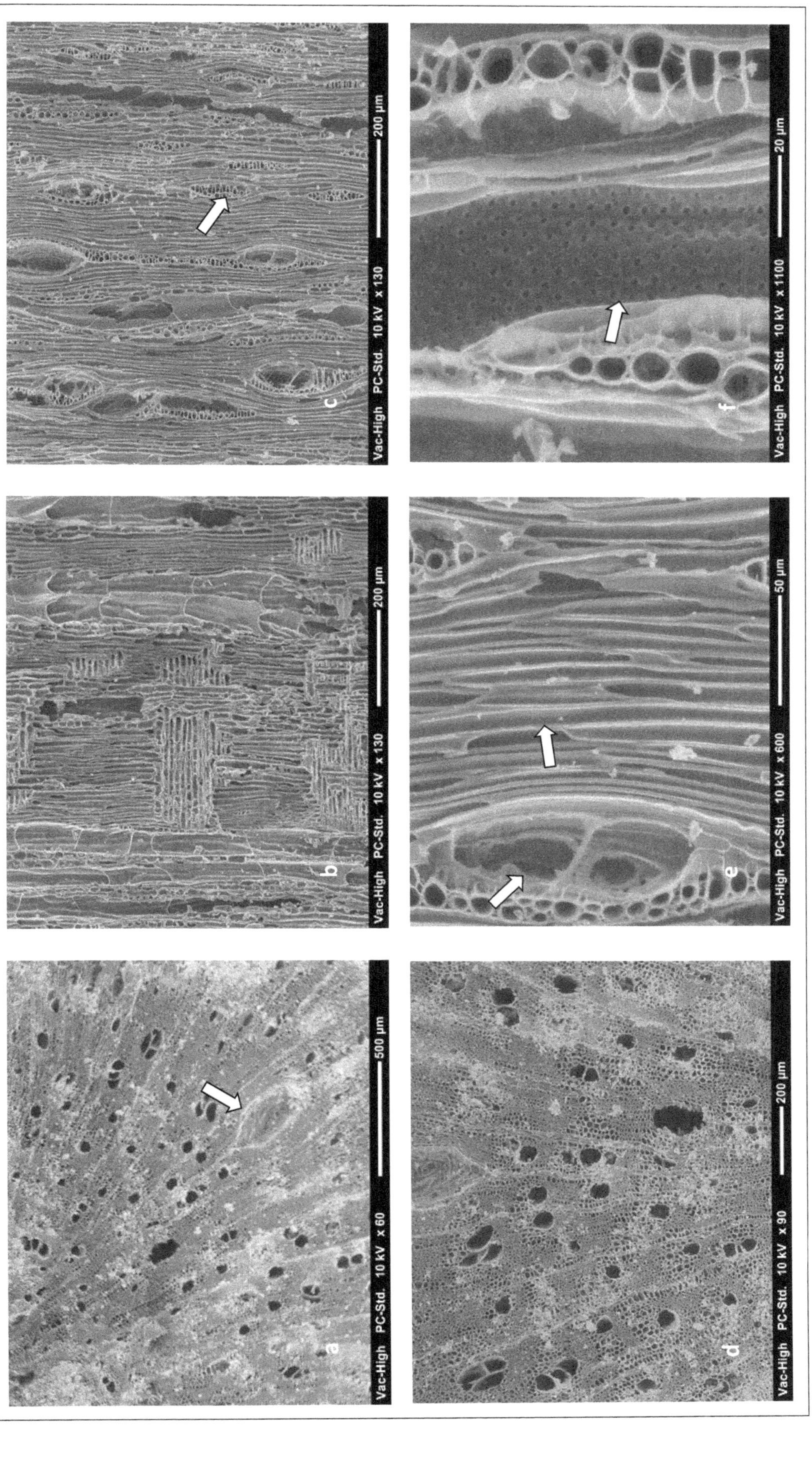

a and d: transversal, b and e: radial, c and f: tangential

A6 - BOMBACACEAE - *Adansonia digitata* L. (Northwestern Botswana)

IAWA Feature Code: 4,9,22,27,41,56,60,65,69,77,112,105,65

- Wood semi-ring porous
- Vessels solitary or in multiples of up to 4
- Simple plate perforations
- Alternate intervessel pitting of 2 sizes
- Intervessel pitting large and round to oval (17µm)
- Intervessel pitting almost scalariform
- Large - >= 10 µm intervessel pits (f)
- Mean tangential diameter vessel lumina: 67µm
- Vascular tracheids present
- Fibres thin to thick walled
- Tyloses common(e)
- Apotracheal parenchyma-parenchyma in aggregates
- Simple pitted ray cells
- Rays 1-3 seriate(c)
- All ray cells upright/and or square
- Axial parenchyma quasi-storied
- Septate fibres present

***Adansonia digitata*. Photo: M. Mvimi**

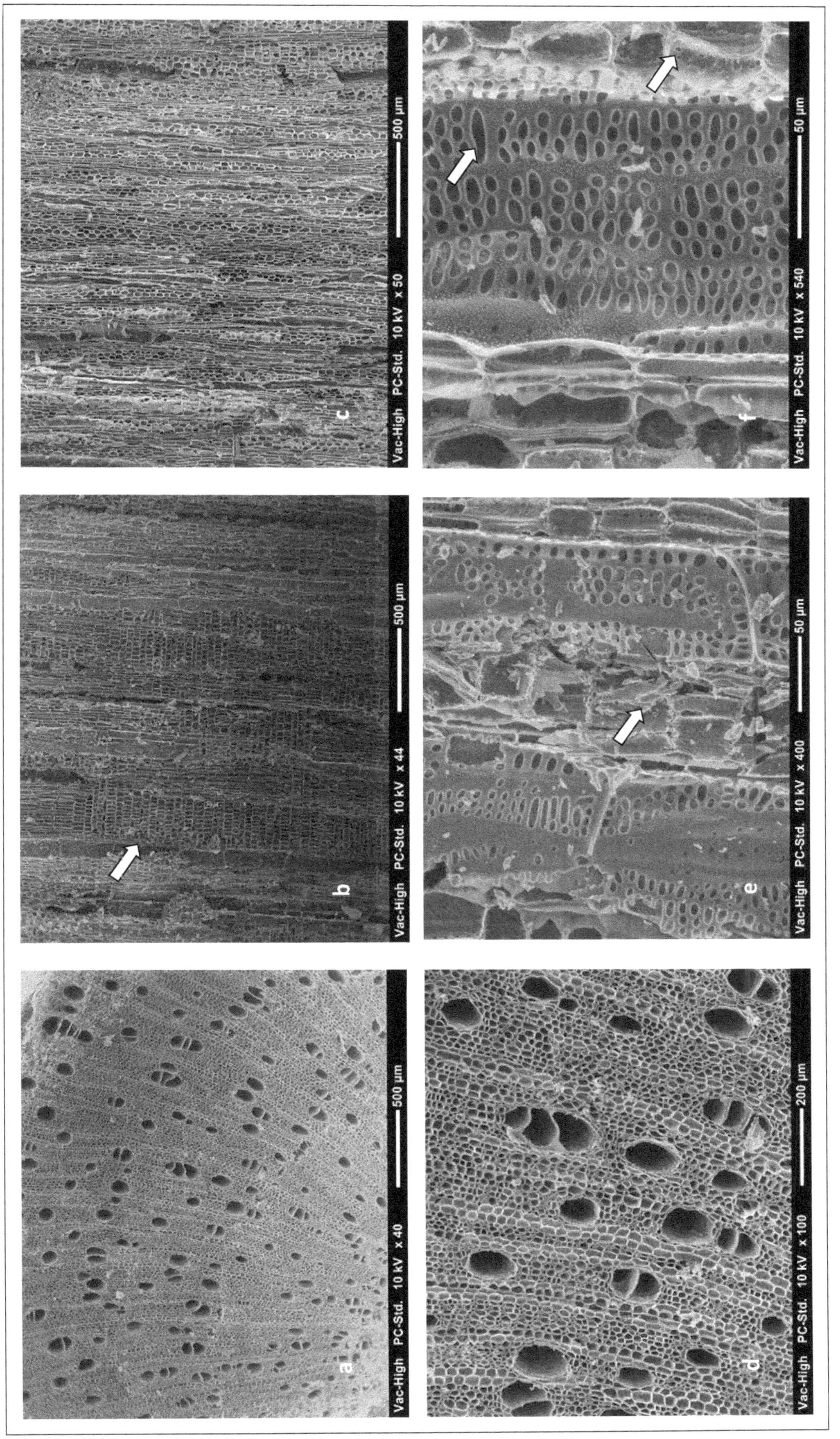

a and d: transversal, b and e: radial, c and f: tangential

A7 – MALVACEAE - *Azanza garckeana* (F. Hoffm.) Exell & Hillcoat (Southeastern Botswana)

IAWA Feature Code: 2,5,10,13,22,50,42,60,65,69,76,97,105,112,124

- Growth rings absent
- Wood diffuse porous
- Vessels solitary, in pairs or in radial multiples of 3
- Simple plate perforations(b)
- Intervessel pitting alternate
- Mineral residues in intervessel pits (medium 7-10µm)
- Vessel-ray pits similar to intervessel pitting
- Helical thickenings absent
- Mean tangential diameter vessel lumina: 100-200µm
- Vasicentric tracheids
- Fibres with thick to thin walls
- Septate fibres present (e)
- Diffuse axial parenchyma
- 1-3 seriate
- All ray cells upright and/or square cells
- Perforated ray cells
- Storied axial parenchyma
- Oil and/mucilage cells in intervessel pits

***Azanza garckeana*. Photo: M. Mvimi**

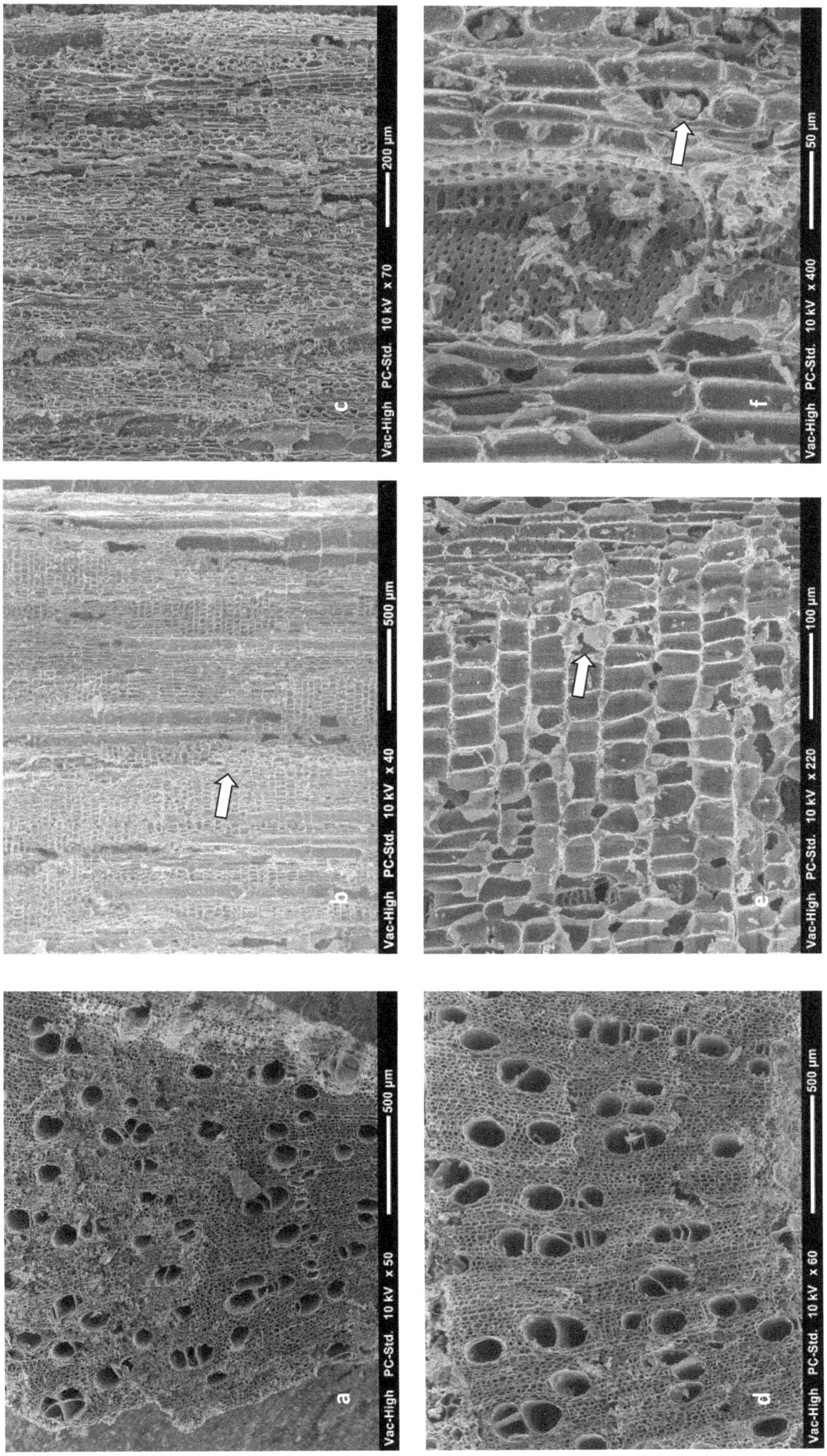

a and d: transversal, b and e: radial, c and f: tangential

A8 – FABACEAE - *Adenolobus garipensis* (E. Mey.) Torre & Hills. (Erongo, Namibia)

IAWA Feature Code: 1,5,8, 11,13,21,30,40,65,69,78,97,103,109,115,136,144

- Growth rings present
- Wood diffuse porous
- Rays in a dendritic arrangement
- Mean tangential diameter vessel lumina: 50µm
- Vessels solitary and also in short and long radial clusters of 2-5
- Simple plate perforations
- Storied ray structure in the transversal side
- Intervessel pits opposite, elongated and quasi-scalariform in some areas (>10µm)
- Intervessel pits of 2 distinct sizes
- Vessel-ray pitting similar to intervessel pitting
- Vessels of 2 distinct diameters (d)
- Fibres medium to thick walled
- Septate fibres present
- Axial parenchyma scanty paratracheal
- 1-2 seriate rays
- A combination of both procumbent and square ray cells
- Rays of two distinct sizes (c)
- Idioblasts present in rays
- Druses in ray cells
- Tyloses in vessels
- Aggregate vessel structures(f)
- Long rays up to 26 cells high
- Shortest rays at ± cells high
- Rays with both uni and bi or triseriate cells present
- Minutely pitted interray cells

***Adenolobus garipensis*. Photo: M. Mvimi**

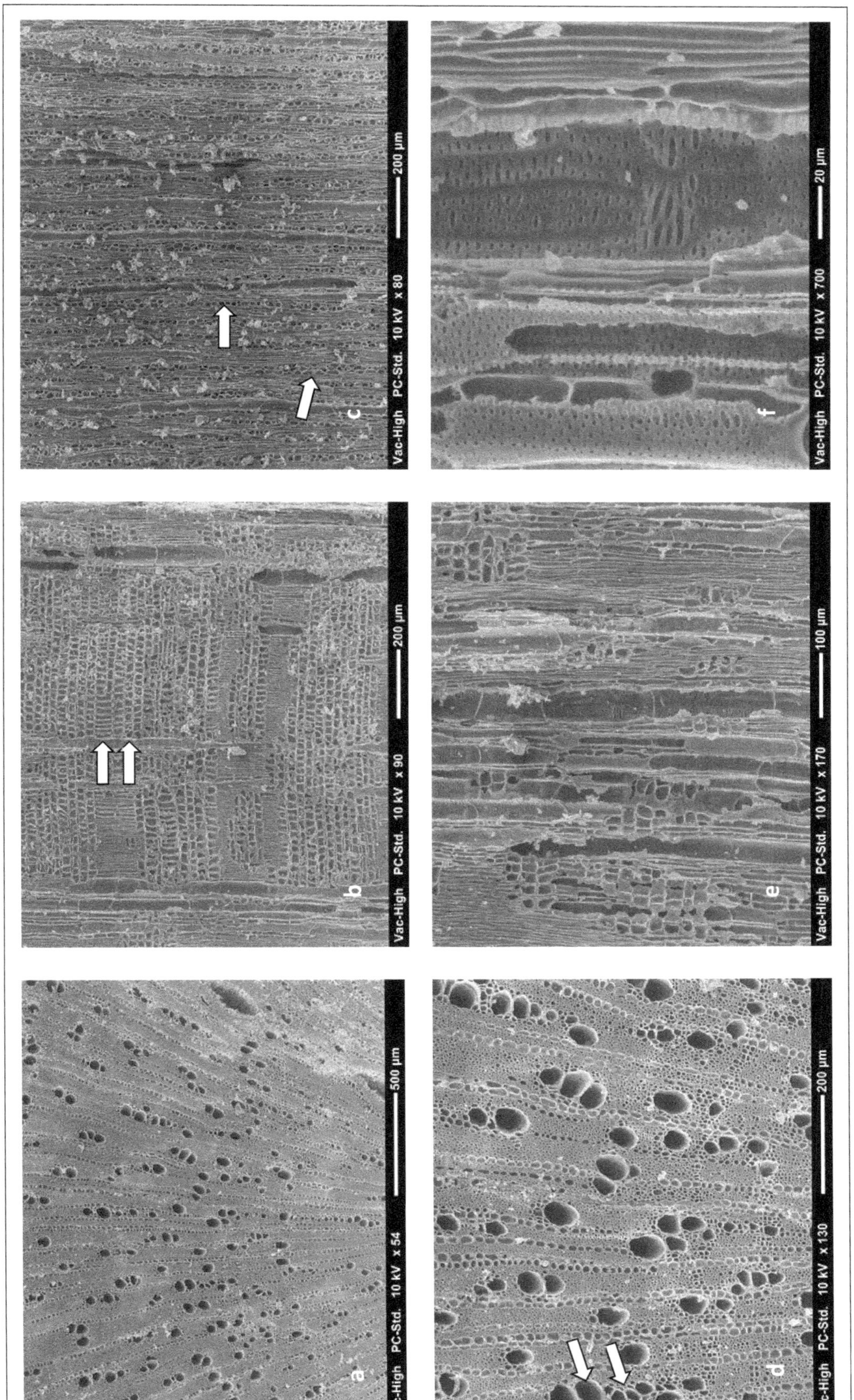

a and d: transversal, b and e: radial, c and f: tangential

A9 – COMBRETACEAE - *Combretum apiculatum* subsp. *boreal* Exell (South-eastern Botswana)

IAWA Feature Code: 1,3,8,9,13,22,30,41,79,85,96,133,136

- Growth ring boundaries distinct
- Wood ring porous
- Vessels in dendritic form
- Vessels < 90% solitary
- Both procumbent and upright ray cells
- Simple perforation plates
- Alternate intervessel pitting (oval shaped)
- Spiral thickenings in fibres
- Vessel-ray pitting similar to intervessel pitting
- Mean tangential diameter vessel lumina: 80μm
- Large crystals in fibres and in ray cells
- Idioblasts in fibre cells
- Crystal diameter: 40μm
- Rays exclusively uniseriate
- Included phloem(d)
- Axial parenchyma paratracheal and vasicentric
- Axial parenchyma banded

Combretum apiculatum. Photo : M. Mvimi

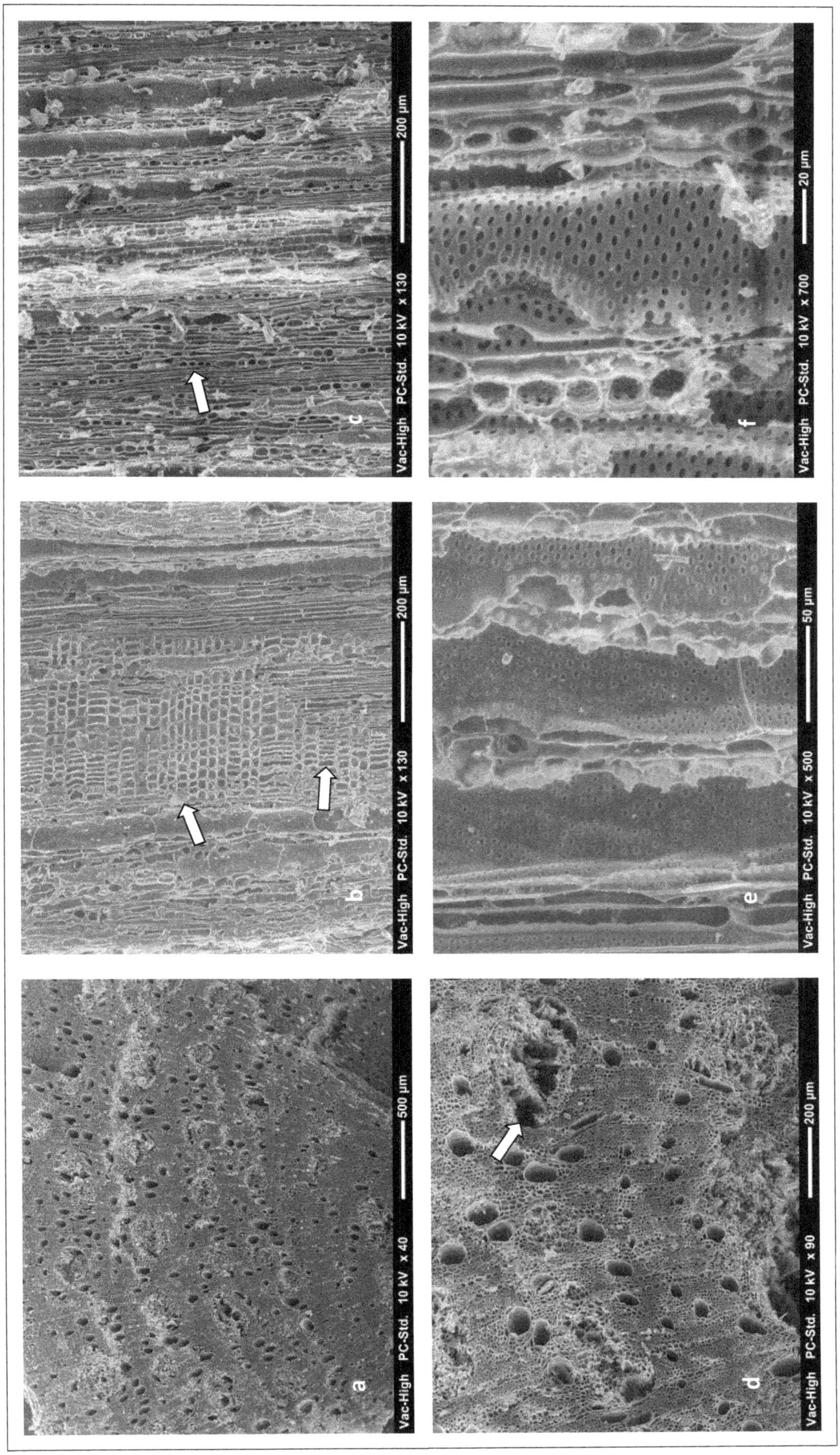
a
Vac-High PC-Std. 10 kV x 40 500 µm
b
Vac-High PC-Std. 10 kV x 130 200 µm
c
Vac-High PC-Std. 10 kV x 130 200 µm
d
Vac-High PC-Std. 10 kV x 90 200 µm
e
Vac-High PC-Std. 10 kV x 500 50 µm
f
Vac-High PC-Std. 10 kV x 700 20 µm

A10 – BURSERACEAE - *Commiphora glandulosa* Schinz (South-eastern Botswana)

IAWA Feature Code: 2,5,9,13,21,25,26,30,41,65,68,76,99,109,144

- Growth ring boundaries absent
- Wood diffuse porous
- Diffuse axial parenchyma
- Vessels rarely solitary, in pairs, or in small radial groups
- Simple plate perforations
- Opposite intervessel pits
- Small to medium intervessel pits
- Intervessel pits 8μm
- Vessel-ray pitting similar to intervessel pitting
- Druses present in rays
- Very thin fibre cells
- Both square and upright ray cells
- Resin canals present
- Rays 1-3 seriate
- Uniseriate rays up to ± 21 rays high
- Shortest rays at 4 cells high
- Minutely pitted parenchyma cells
- Mean tangential diameter vessel lumina: 67μm
- Septate fibres present

Commiphora glandulosa. **Photo M. Mvimi**

a: transversal, b and e: radial, c, d and f: tangential

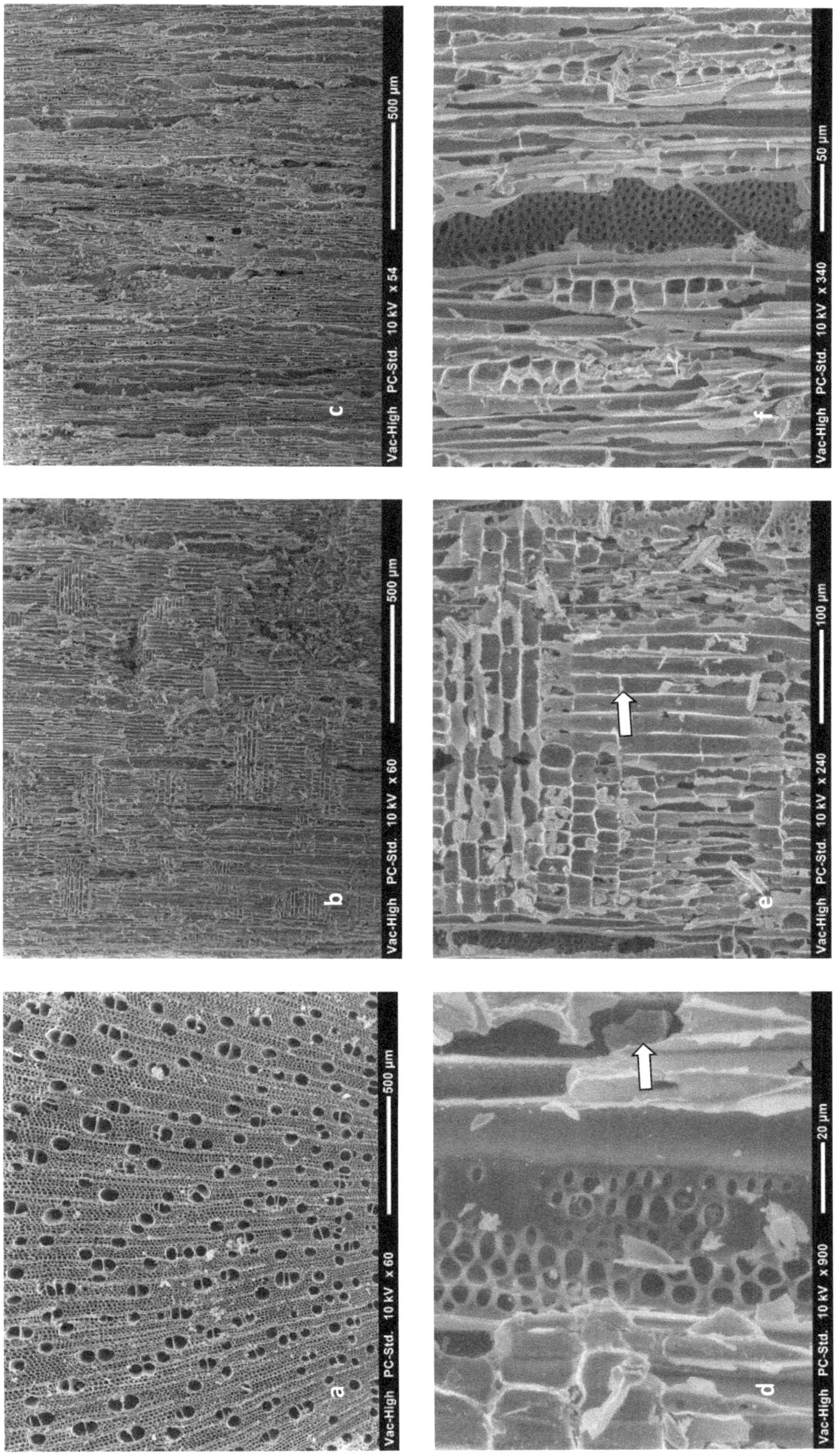

A11 – BURSERACEAE - *Commiphora mossambicensis* Engl. (North-western Botswana)

IAWA Feature Code: 2,5,13,27,36,41,65,68,97,102,109,130,144

- Growth ring boundaries absent
- Wood diffuse porous
- 1-2 seriate
- Simple perforation plates
- Mean tangential diameter vessel lumina: 71μm
- Druses in ray cells
- Crystals present in ray cells
- Large intervessel pits ≤ 11μm
- Highest rays at 13 cells high
- Shortest rays at 3 cells high
- Uniseriate rays generally shorter than multiseriate rays
- Cupresoid shaped intervessel pits
- Very thin fibre cells
- Septate fibres present
- Both upright and procumbent ray cells
- Minutely simple pitted parenchyma cells
- Resin canals present

Commiphora mossambicensis. **Photo: G. Tabushwa**

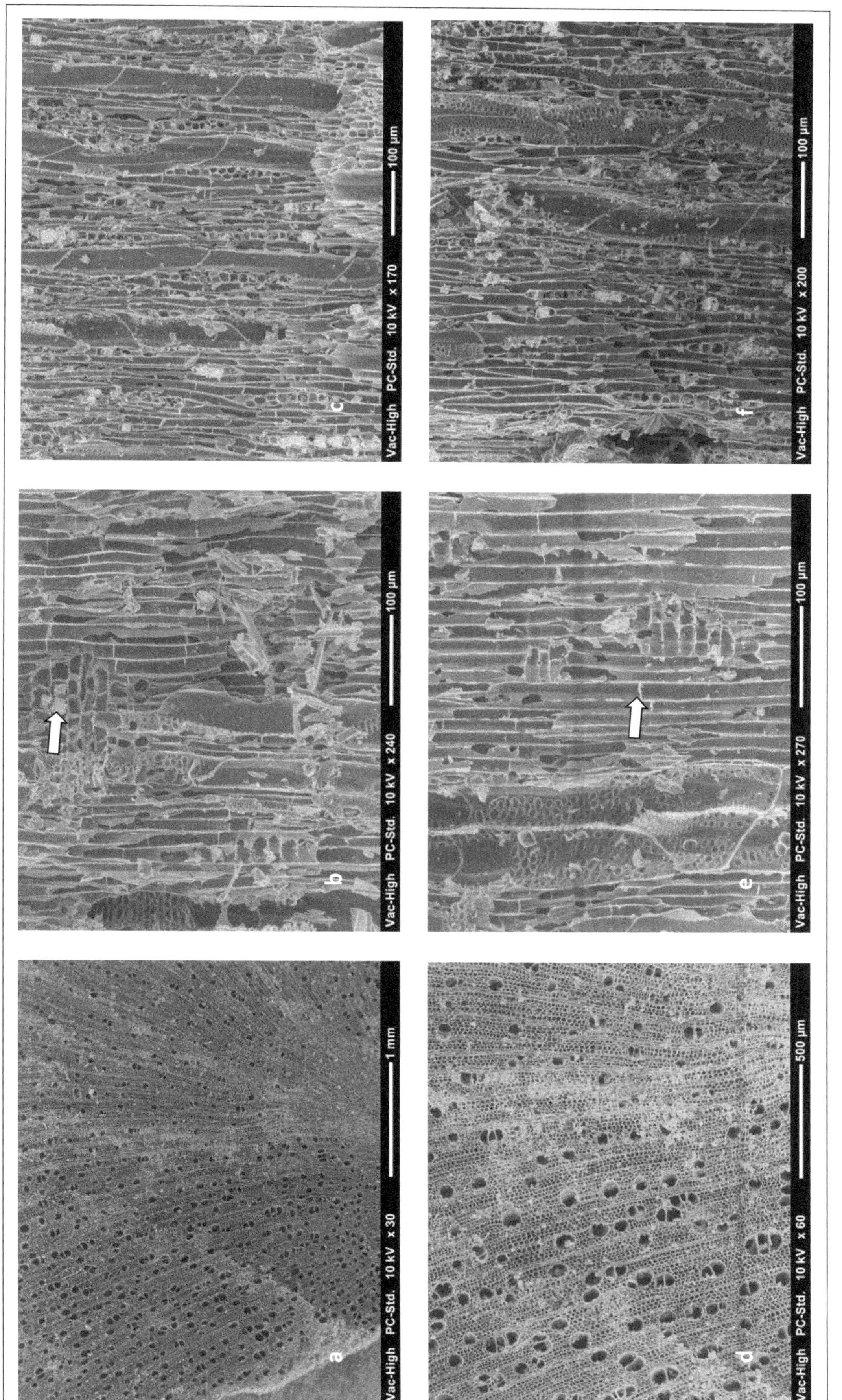

a and d: transversal, b and e: radial, c and f: tangential

A12 – FABACEAE - *Dichrostachys cinerea* (L.) Wight & Arn (Erongo, Namibia)

IAWA Feature Code: 2,5,9,10, 13,22,24,41,47,53,56,61,65,79,83,97,102,104,115,136,142

- Growth rings present
- Wood diffuse porous
- Rays in a dendritic arrangement
- Mean tangential diameter vessel lumina: 50µm
- Vessels solitary and also in short and long radial clusters of 2-5
- Simple plate perforations
- Storied ray structure in the transversal side
- Intervessel pits opposite, elongated and quasi-scalariform in some areas (>10µm)
- Intervessel pits of 2 distinct sizes
- Vessel-ray pitting similar to intervessel pitting
- Vessels of 2 distinct diameters
- Fibres medium to thick walled
- Septate fibres present
- Axial parenchyma scanty paratracheal
- 1-2 seriate rays
- A combination of both procumbent and square ray cells
- Idioblasts present in rays
- Druses in ray cells
- Tyloses in vessels
- Aggregate vessel structures
- Long rays up to 26 cells high
- Shortest rays at ± cells high
- Rays with both uni and bi or triseriate cells present
- Minutely pitted interray cells

***Dichrostachys cinerea*. Photo: M. Mvimi**

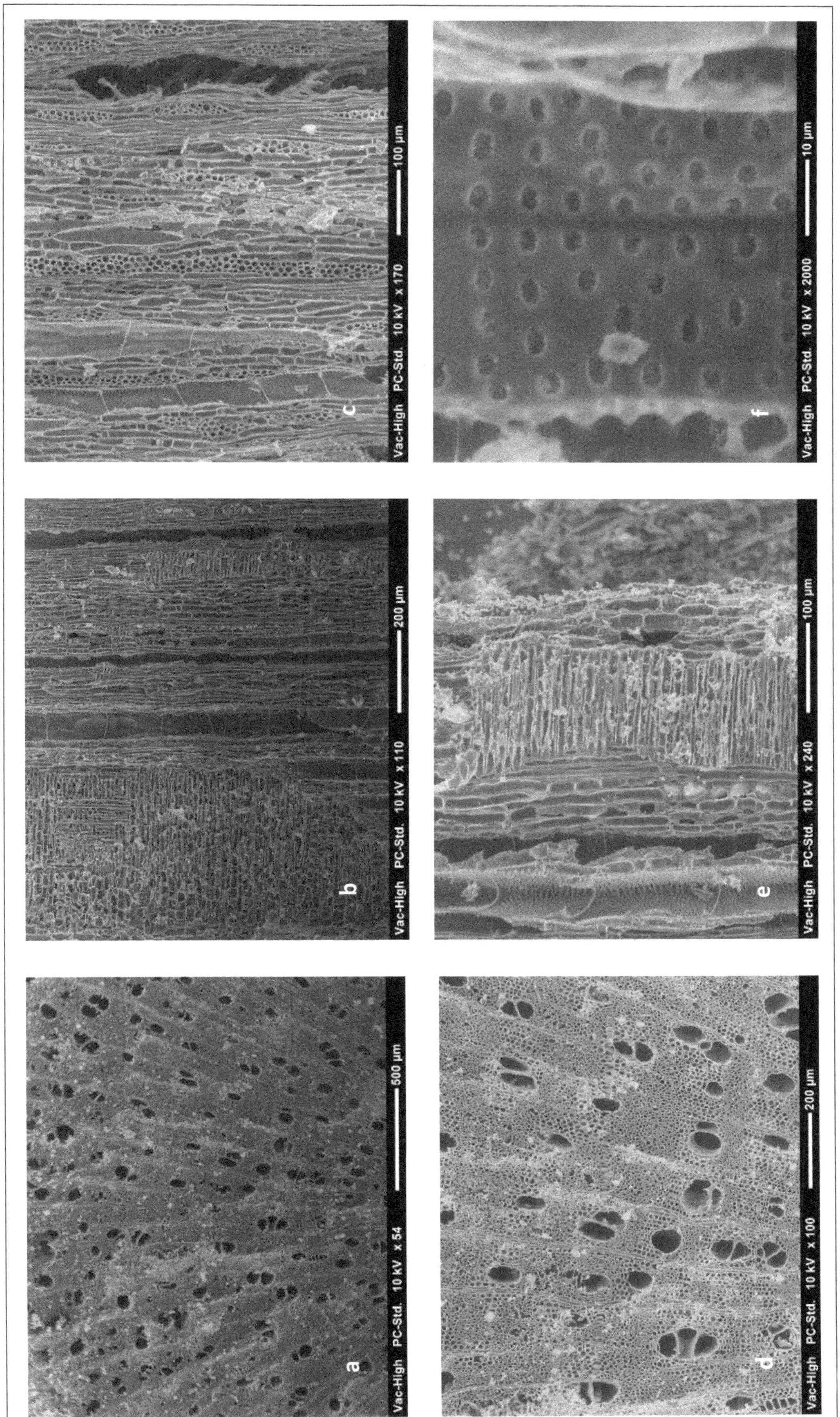
a
Vac-High PC-Std. 10 kV x 54
500 µm
b
Vac-High PC-Std. 10 kV x 110
200 µm
c
Vac-High PC-Std. 10 kV x 170
100 µm
d
Vac-High PC-Std. 10 kV x 100
200 µm
e
Vac-High PC-Std. 10 kV x 240
100 µm
f
Vac-High PC-Std. 10 kV x 2000
10 µm

A13 - MALVACEAE - *Dombeya rotundifolia* Planch. (South-eastern Botswana)

IAWA Feature Code: 2,5,8,10,13,26,50,52,65,70,78,97,108,114

- Growth ring boundaries indistinct or absent
- Wood diffuse-porous
- Vessels in dendritic pattern
- Vessels in radial multiples of 4 or more common
- Solitary vessel outlines circular
- Mean tangential diameter vessel lumina: <= 50 µm
- Simple plate perforations (f)
- Intervessel pits of 2 distinct sizes: opposite, and quasi-scalariform
- Intervessel pits medium: 7-10µm
- >= 100 vessels per square millimetre
- Mean vessel element length: <= 350 µm
- Septate fibres present
- Fibres very thick-walled
- Axial parenchyma scanty paratracheal
- Ray width 1 to 3 cells
- Ray height > 1 mm
- Body ray cells procumbent with over 4 rows of upright and/or square marginal cells
- <= 4 / mm rays per millimetre

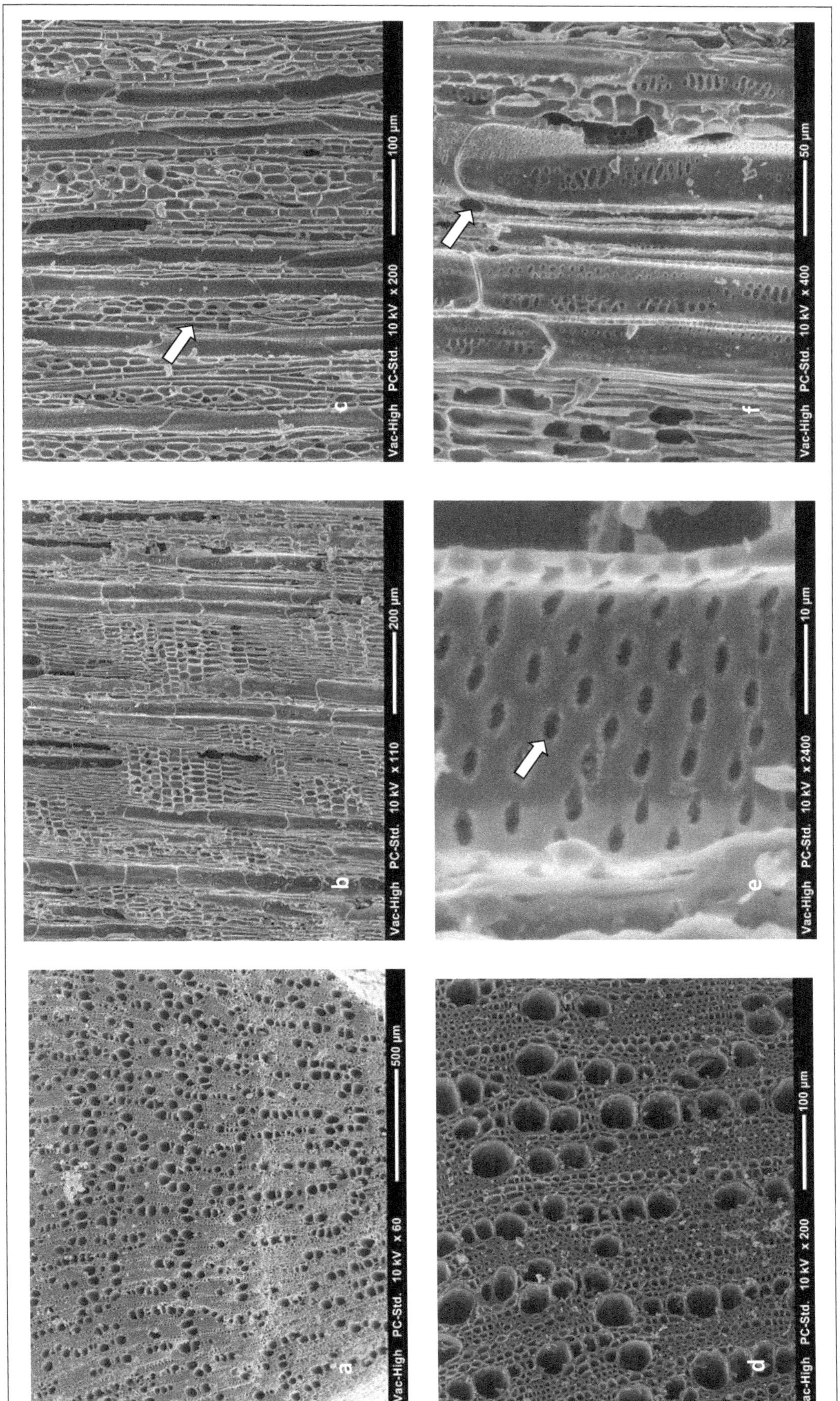
a
Vac-High PC-Std. 10 kV x 60
500 µm
b
Vac-High PC-Std. 10 kV x 110
200 µm
c
Vac-High PC-Std. 10 kV x 200
100 µm
d
Vac-High PC-Std. 10 kV x 200
100 µm
e
Vac-High PC-Std. 10 kV x 2400
10 µm
f
Vac-High PC-Std. 10 kV x 400
50 µm

A14 – CAPPARACEAE - *Boscia foetida* Schinz. (Erongo, Namibia)

IAWA Feature Code: 1,4,10,11,13,22,24,41,50,52,58,61,65,69,85,86,97,102,104,115,137,144

- Growth ring boundaries distinct
- Vessels in dendritic pattern
- Vessel clusters common
- Vessels in radial multiples of 4 or more common
- Simple perforation plates (f)
- Intervessel pits alternate
- Intervessel pits minute: <=4 µm
- Mean tangential diameter of vessel lumina: 50-100µm
- Gums and other deposits in heartwood vessels
- Fibres with simple to minutely bordered pits
- Septate fibres present
- Fibres thin- to thick-walled
- Axial parenchyma bands more than three cells wide
- Axial parenchyma in narrow bands or lines up to three cells wide
- Ray width 1 to 2 cells
- Ray height: >1mm
- All ray cells procumbent
- Rays: 4-12 / mm
- >= 100 vessels per square millimetre
- Mean vessel element length: <= 350µm
- Prismatic crystals in upright and/or square ray cells
- Druses present

***Boscia foetida*. Photo: M. Mvimi**

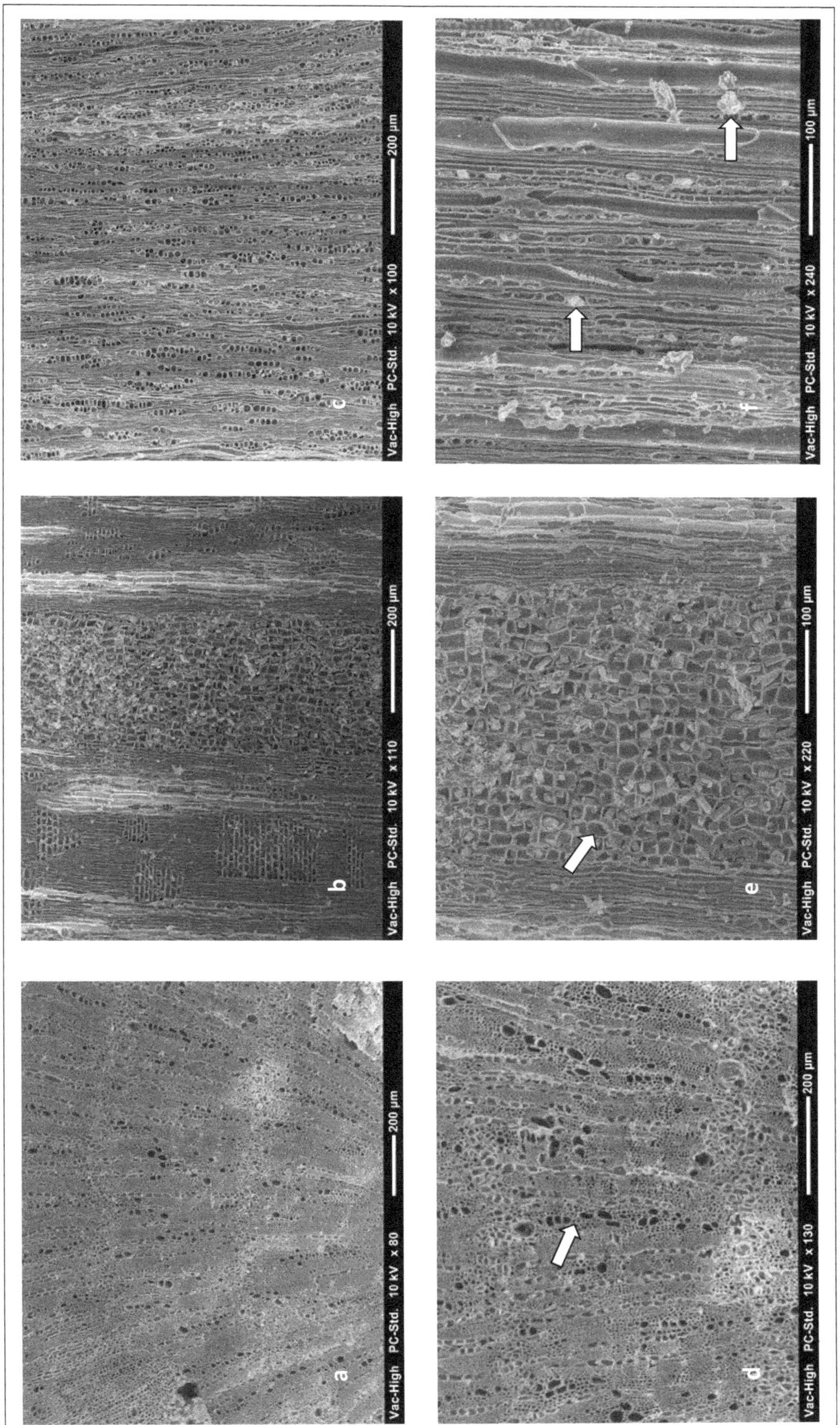
a
Vac-High PC-Std. 10 kV x 80
200 µm
b
Vac-High PC-Std. 10 kV x 110
200 µm
c
Vac-High PC-Std. 10 kV x 100
200 µm
d
Vac-High PC-Std. 10 kV x 130
200 µm
e
Vac-High PC-Std. 10 kV x 220
100 µm
f
Vac-High PC-Std. 10 kV x 240
100 µm

A15 – RUBIACEAE - *Gardenia volkensii* K. Schum (South-eastern Botswana)

IAWA Feature Code: 2,5,9,13,22,24,41,47,58,61,65,68,96,133,137,144

- Growth ring boundaries indistinct or absent
- Vessels exclusively solitary (90% or more)
- Simple perforation plates (f)
- Intervessel pits alternate
- Intervessel pits minute: <=4 µm
- Mean tangential diameter of vessel lumina: 50-100µm
- Gums and other deposits in vessels and fibres
- Fibres with simple to minutely bordered pits
- Septate fibres absent
- Fibres thin-walled
- Axial parenchyma absent or extremely rare
- Rays exclusively uniseriate
- Ray height: >1mm
- Body ray cells procumbent with over 4 rows of upright and/or square marginal cells
- Rays: <=4-12 / mm
- 5-20 vessels per square millimetre
- Mean vessel element length: <= 350µm
- Prismatic crystals in upright and/or square ray cells
- Druses present
- Included phloem

***Gardenia volkensii.* Photo: M. Mvimi**

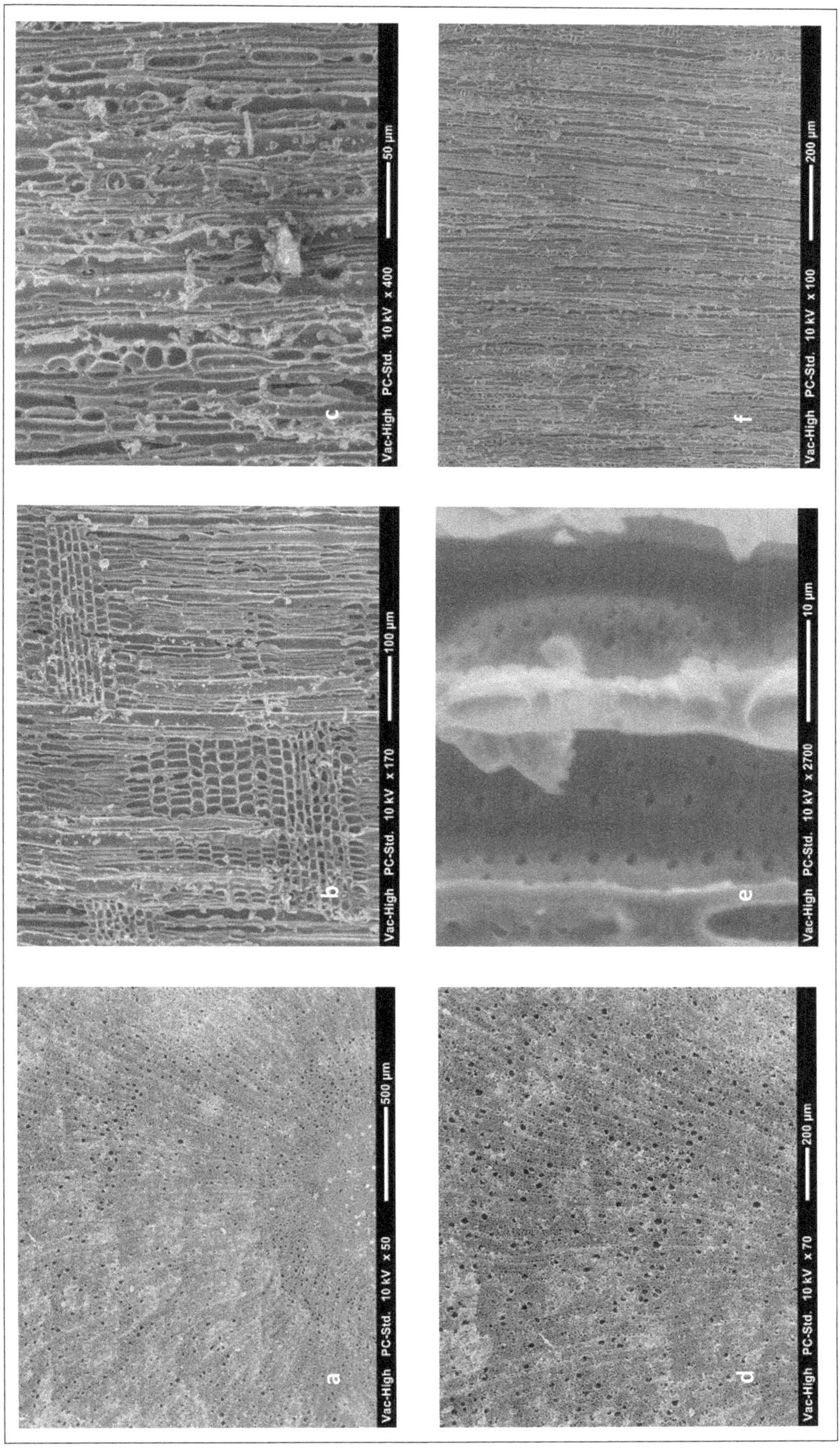

a and d: transversal, b and e: radial, c and f: tangential

A16 – TILIACEAE - *Grewia flava* DC. (South-eastern Botswana)

IAWA Feature Code: 1,3,11,13,22,25,41,50,52,58,69,75,97,102,109,115

- Growth ring boundaries distinct (a)
- Vessel clusters common
- Simple perforation plates (f)
- Intervessel pits alternate
- Intervessel pits small - 4 - 7 µm
- Mean tangential diameter of vessel lumina: 50-100µm
- Gums and other deposits in heartwood vessels
- Fibres thin-to thick-walled
- Axial parenchyma absent or extremely rare
- Ray width 1 to 3 cells
- Ray height: >1mm
- Rays with procumbent, square and upright cells mixed throughout the ray
- Rays: 4-12 / mm
- >= 100 vessels per square millimetre
- Mean vessel element length: <= 350µm

***Grewia flava*. Photo: M. Mvimi**

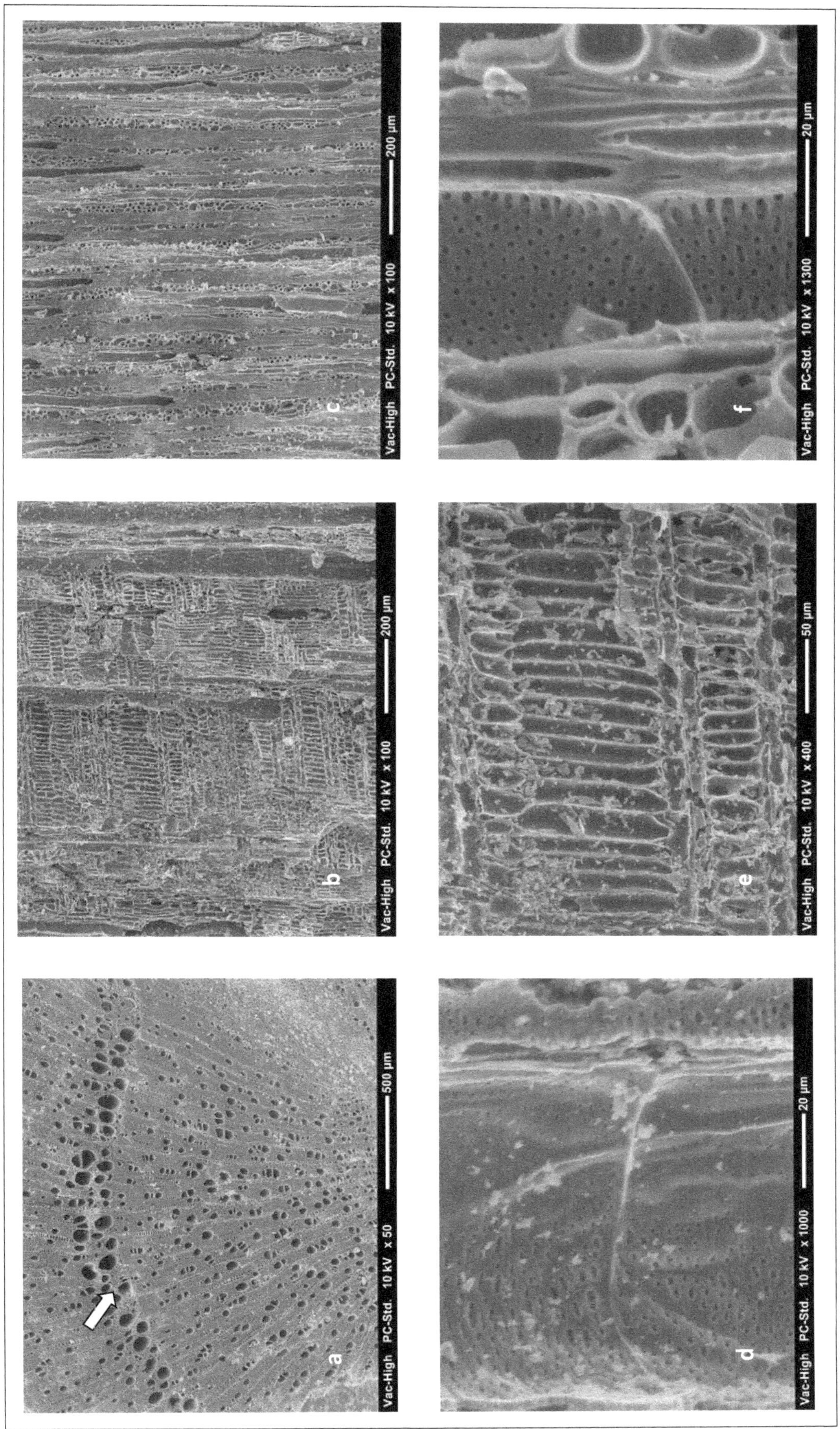

a: transversal, b and e: radial, c, d and f: tangential

A17 – MORINGACEAE - *Moringa ovalifolia* Denter&Berger (Erongo mountains)

IAWA Feature Code: 2,9,13,22,24,41,47,54,58,61,78,96,102,108,115,144

- Growth ring boundaries indistinct or absent
- Vessels exclusively solitary (90% or more)
- Simple perforation plates
- Intervessel pits alternate
- Intervessel pits minute: <=4 µm
- Mean tangential diameter of vessel lumina: 50-100µm
- Gums and other deposits in parenchyma and fibre cells
- Fibres with simple to minutely bordered pits
- Fibres thin-walled
- Axial parenchyma absent or extremely rare
- Rays exclusively uniseriate
- Ray height: >1mm
- Body ray cells procumbent with over 4 rows of upright and/or square marginal cells
- Rays: <=4-12 / mm
- 5-20 vessels per square millimetre
- Mean vessel element length: <= 350µm
- Druses present

Moringa ovalifolia*. Photo: *www.africanplants.senckenberg.de

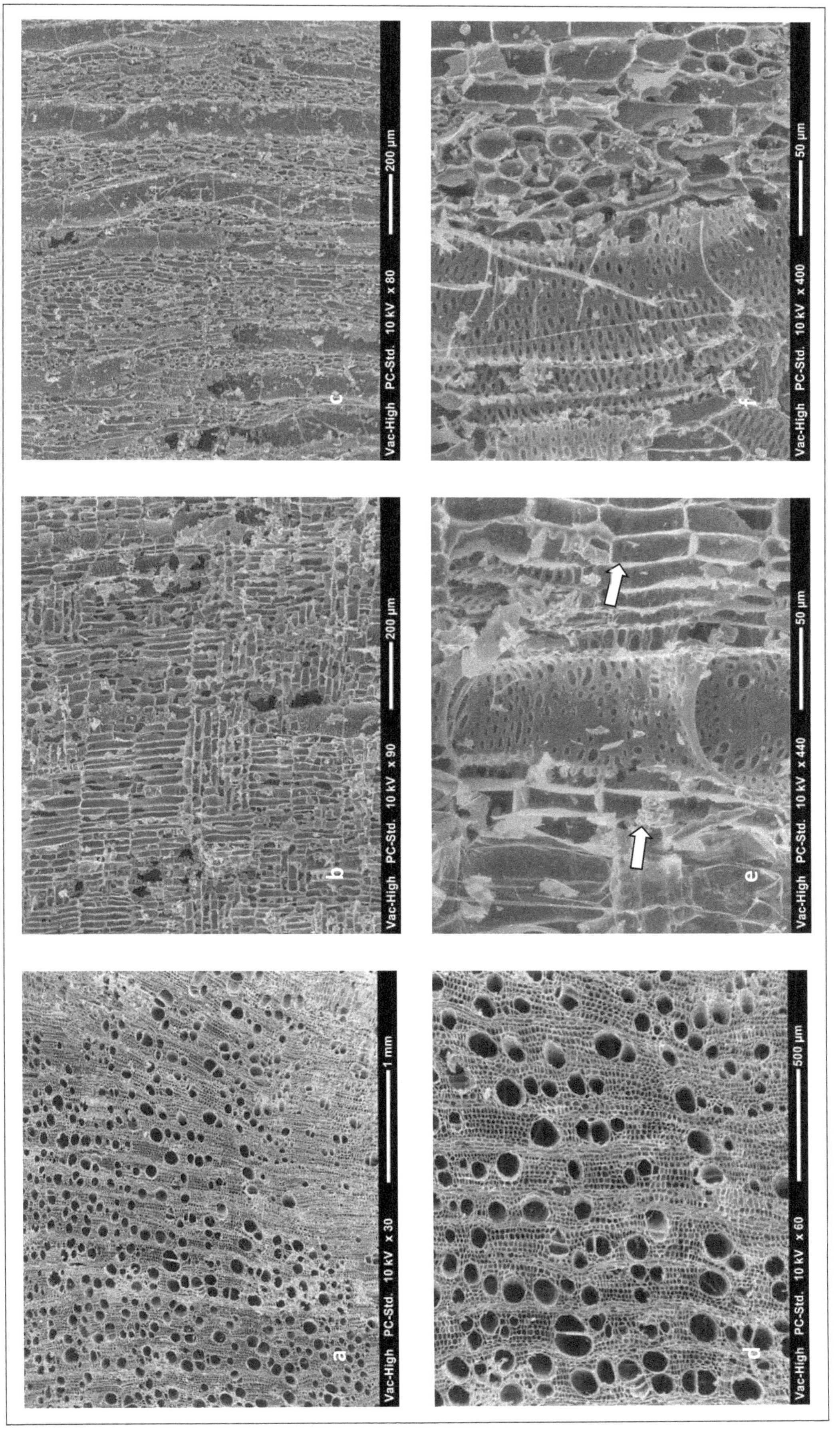

a and d: transversal, b and e: radial, c and f: tangential

A18 – SAPOTACEAE - *Sideroxylon inerme* L. (South-eastern Botswana)

IAWA Feature Code: 2,11,14,41,47,58,65,68,76,85,92,85,92,97,109,115,144

IAWA Feature Code: 2,11,14,41,47,58,65,68,76,85,92,85,92,97,109,115,144

- Growth ring boundaries indistinct or absent
- Nested-vessel pattern
- Vessel clusters common
- Simple perforation plates
- Intervessel pits scalariform
- Mean tangential diameter of vessel lumina: 50-100μm
- Mean tangential diameter of nested vessels: 100μm
- 5-20 vessels per square millimetre
- Fibres thin-walled
- Axial parenchyma diffuse
- Axial parenchyma bands more than three cells wide
- Four (3-4) cells per parenchyma strand
- Banded axial parenchyma present
- Ray width 1 to 3 cells
- Rays of two distinct sizes (c and f)
- Rays with procumbent, square and upright cells mixed throughout the ray
- Rays: 4-12 / mm
- Septate fibres absent
- Druses present
- Gums and other deposits in parenchyma and fibre cells

***Sideroxylon inerme*. Photo : M. Mvimi**

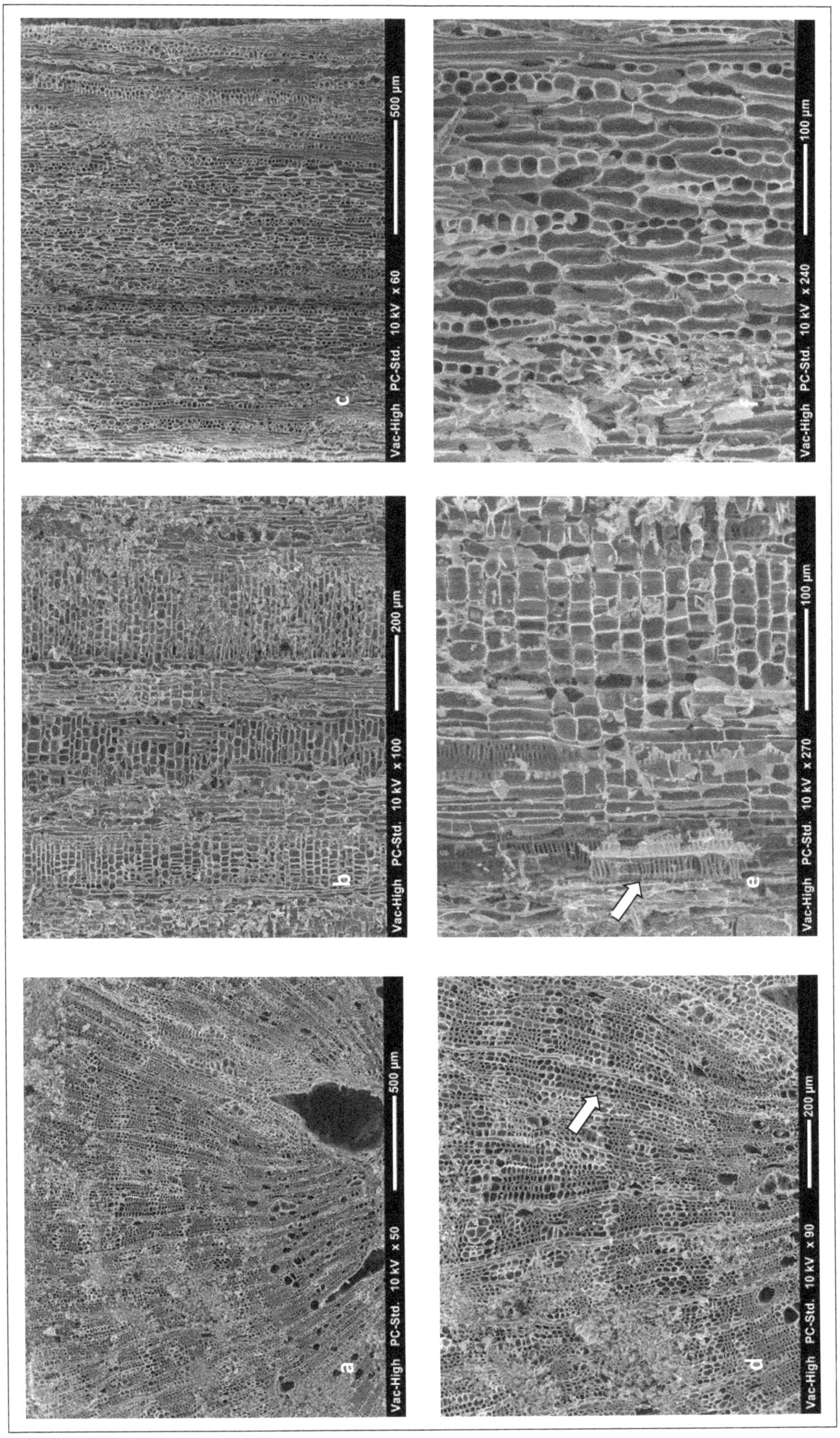
a
Vac-High PC-Std. 10 kV x 50
500 µm
b
Vac-High PC-Std. 10 kV x 100
200 µm
c
Vac-High PC-Std. 10 kV x 60
500 µm
d
Vac-High PC-Std. 10 kV x 90
200 µm
e
Vac-High PC-Std. 10 kV x 270
100 µm
Vac-High PC-Std. 10 kV x 240
100 µm

A19 – CAESALPINIACEAE - *Colophospermum mopane* (J. Kirk ex Benth.) J. Léonard (North-western Botswana)

IAWA Feature Code: 2,5,9,13,22,30,41,48,58, 70,78,96,102,104,112,144

- Growth ring boundaries indistinct or absent
- Wood diffuse porous
- Vessels exclusively solitary (90% or more)
- Solitary vessel outlines circular
- Simple perforation plates
- Intervessel pits alternate
- Mean tangential diameter of vessel lumina: 50-100µm
- 20-40 vessels per square millimetre
- Fibres very thick-walled (d)
- Axial parenchyma scanty paratracheal
- Ray exclusively uniseriate
- Ray height: <1mm
- Ray cells procumbent
- Perforated ray cells
- Gums and other deposits in vessel elements
- Rays: <=4-12 / mm
- Septate fibres present
- Druses present

Colophospermum mopane. **Photo: D. Pleurdeau**

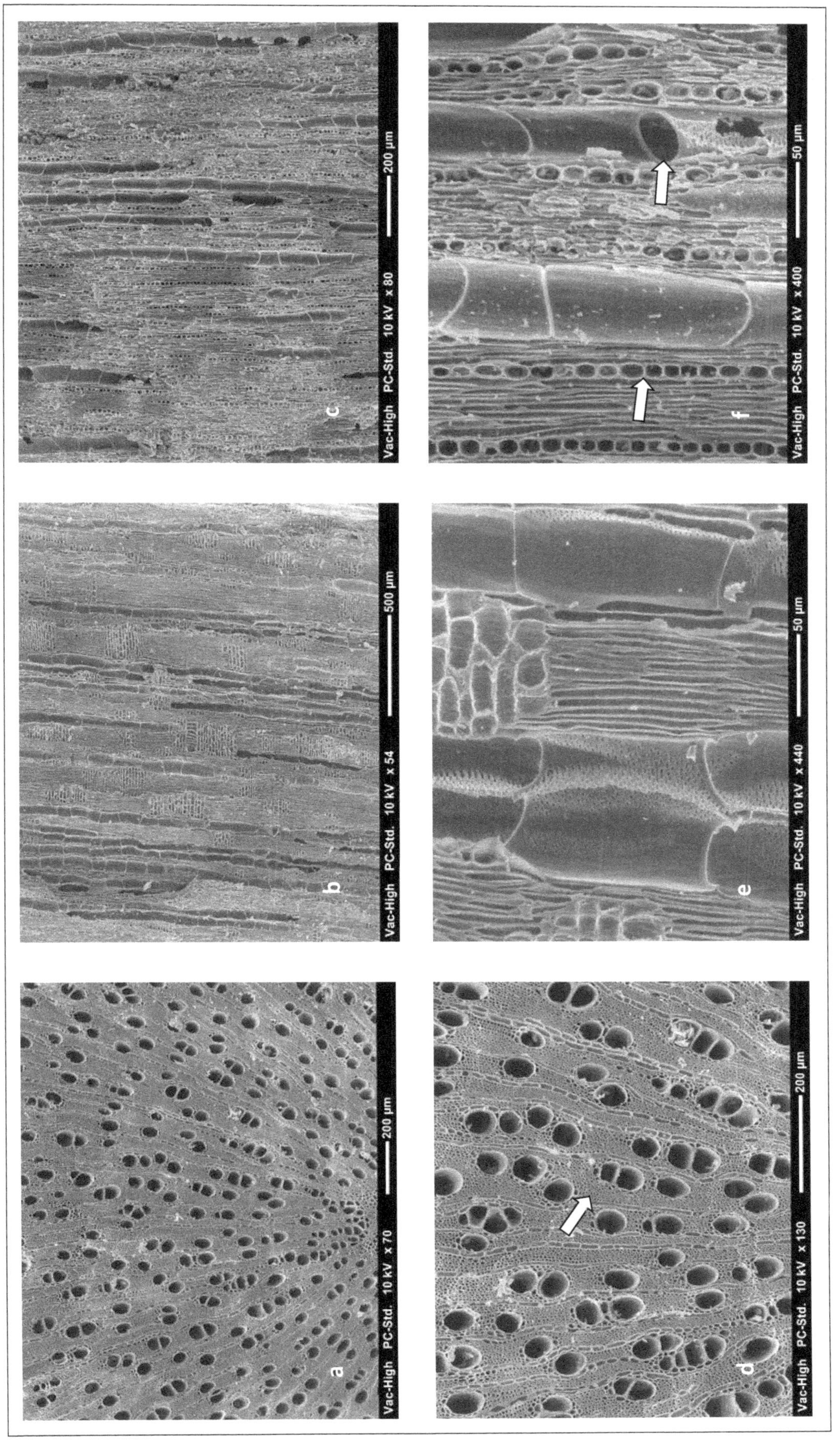
a
Vac-High PC-Std. 10 kV x 70
200 µm
b
Vac-High PC-Std. 10 kV x 54
500 µm
c
Vac-High PC-Std. 10 kV x 80
200 µm
d
Vac-High PC-Std. 10 kV x 130
200 µm
e
Vac-High PC-Std. 10 kV x 440
50 µm
f
Vac-High PC-Std. 10 kV x 400
50 µm

A20 – VITACEAE - *Cissus cornifolia* Planch. (North-western Botswana)

IAWA Feature Code: 1,3,9,13,22,27,47,68,76,92,97,102,105,112,115,144

- Growth ring boundaries distinct
- Wood ring porous
- Vessels exclusively solitary (90% or more)
- Solitary vessel outlines circular
- Simple perforation plates
- Intervessel pits alternate
- Gash-like intervessel pits
- Mean tangential diameter of vessel lumina: 100-200µm
- 5-20 vessels per square millimetre
- Fibres thin-walled
- Axial parenchyma diffuse
- Four (3-4) cells per parenchyma strand
- Ray width 1 to 3 cells
- Ray height: >1mm
- Ray cells upright(e)
- Perforated ray cells
- Rays: 4-12 / mm
- Druses present in fibre cells

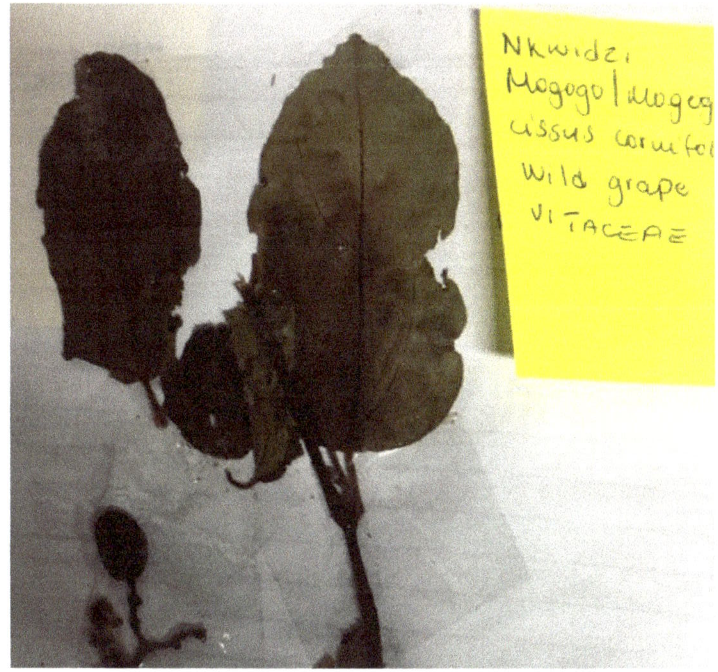

***Cissus cornifolia*. Photo: M. Mvimi**

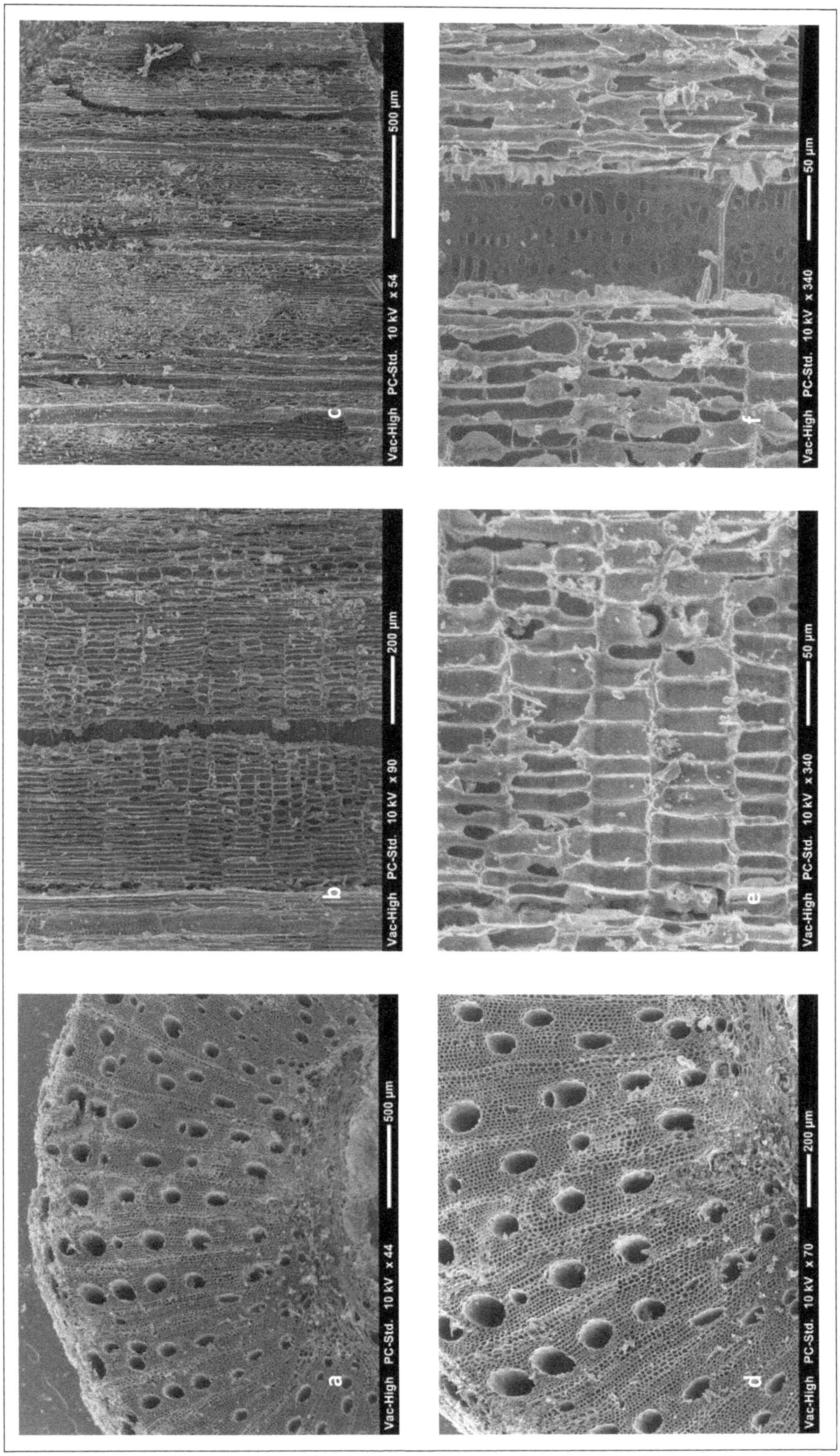

a and d: transversal, b and e: radial, c and f: tangential

A21 – ANACARDIACEAE - *Lannea schweinfurthii* Engl. (South-eastern Botswana)

IAWA Feature Code: 2,3,9,13,16,22,27,41,47,68,76,92,97,102,109

- Growth ring boundaries distinct
- Wood ring porous
- Vessels exclusively solitary (90% or more)
- Vessel clusters present
- Solitary vessel outlines circular
- Perforation plates simple
- Intervessel pits scalariform
- Scalariform perforation plates with 10 - 20 bars
- Mean tangential diameter of vessel lumina: 100-200μm
- 5-20 vessels per square millimetre
- Fibres thin-walled
- Axial parenchyma absent or extremely rare
- Rays exclusively uniseriate
- Ray height: >1mm
- Rays with procumbent, square and upright cells mixed throughout the ray
- Perforated fibre cells
- Rays: <=4-12 / mm
- Prismatic crystals present
- Druses present in fibre cells

Lannea schweinfurthii. **Photo: M. Mvimi**

a and d: transversal, b and e: radial, c and f: tangential

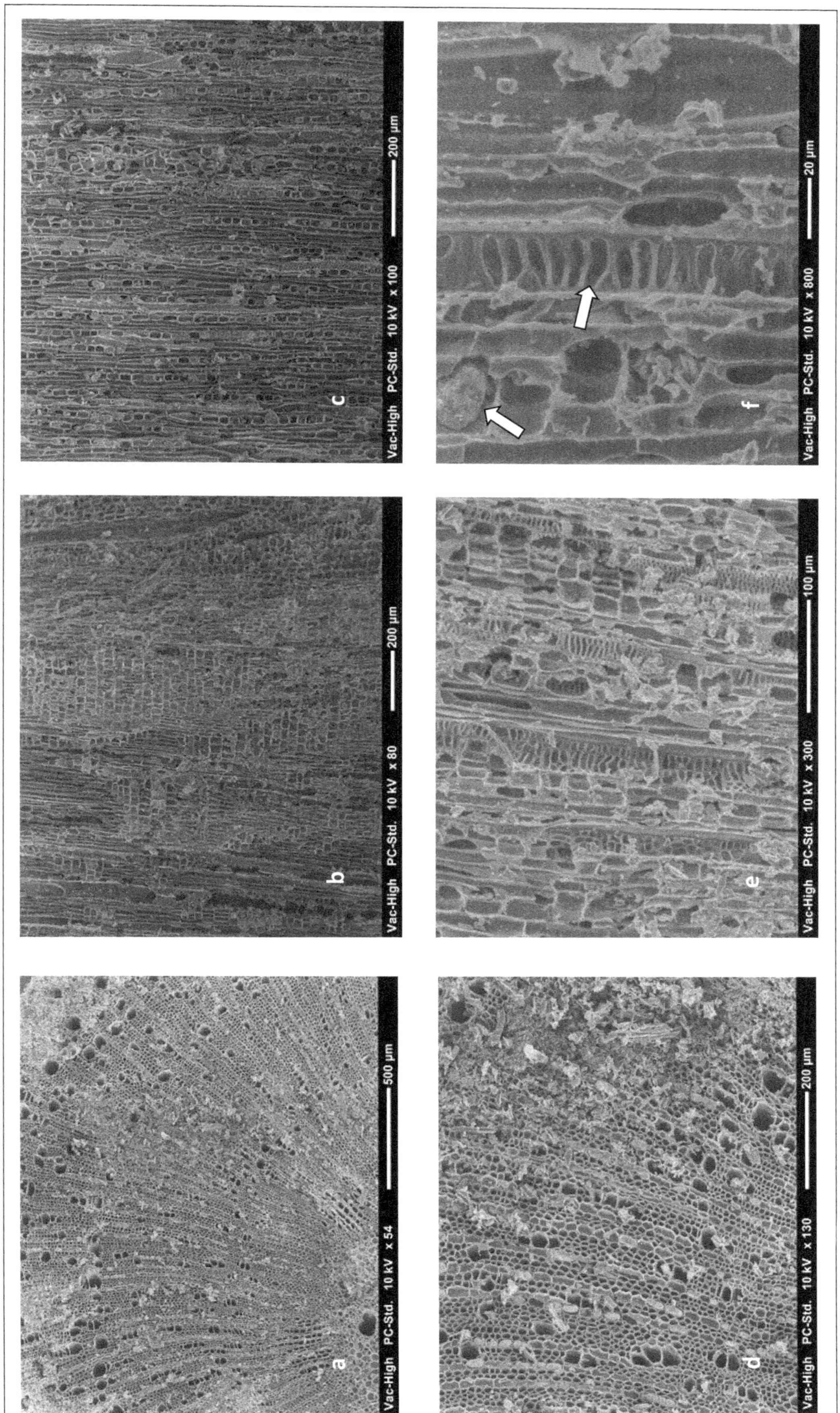

A22 – COMBRETACEAE - *Terminalia prunoides* M.A Lawson (North-western Botswana)

IAWA Feature Code: 2,5,9,10, 13,22,24,41,47,53,56,61,65,79,83,97,102,104,115,136,142

- Growth ring boundaries indistinct or absent
- Wood semi-ring porous
- Vessels mostly solitary and rarely in pairs or radial multiples
- Solitary vessel outline rounded
- Simple perforation plates
- Alternate intervessel pitting
- Minute intervessel pits: <= 4µm
- Mean tangential diameter of vessel lumina: 50-100µm
- Gums and other deposits in vessels
- >=100 vessels per square millimetre
- Fibres thick-walled
- Axial parenchyma diffuse
- Minute alternate pitting in parenchyma cells: <= 4µm
- Rays 1-2 cells wide (b and c)
- Ray height: >1mm
- Rays with procumbent, square and upright cells mixed throughout the ray
- 4-12/ rays per millimetre
- Gums and other deposits in ray cells
- Gums and other deposits in vessels (a)
- Druses in ray parenchyma cells
- Crystal idioblasts present in ray-parenchyma cells (e)
- Prismatic crystals in chambered axial parenchyma cells

Terminalia prunoides **Photo : M. Mvimi**

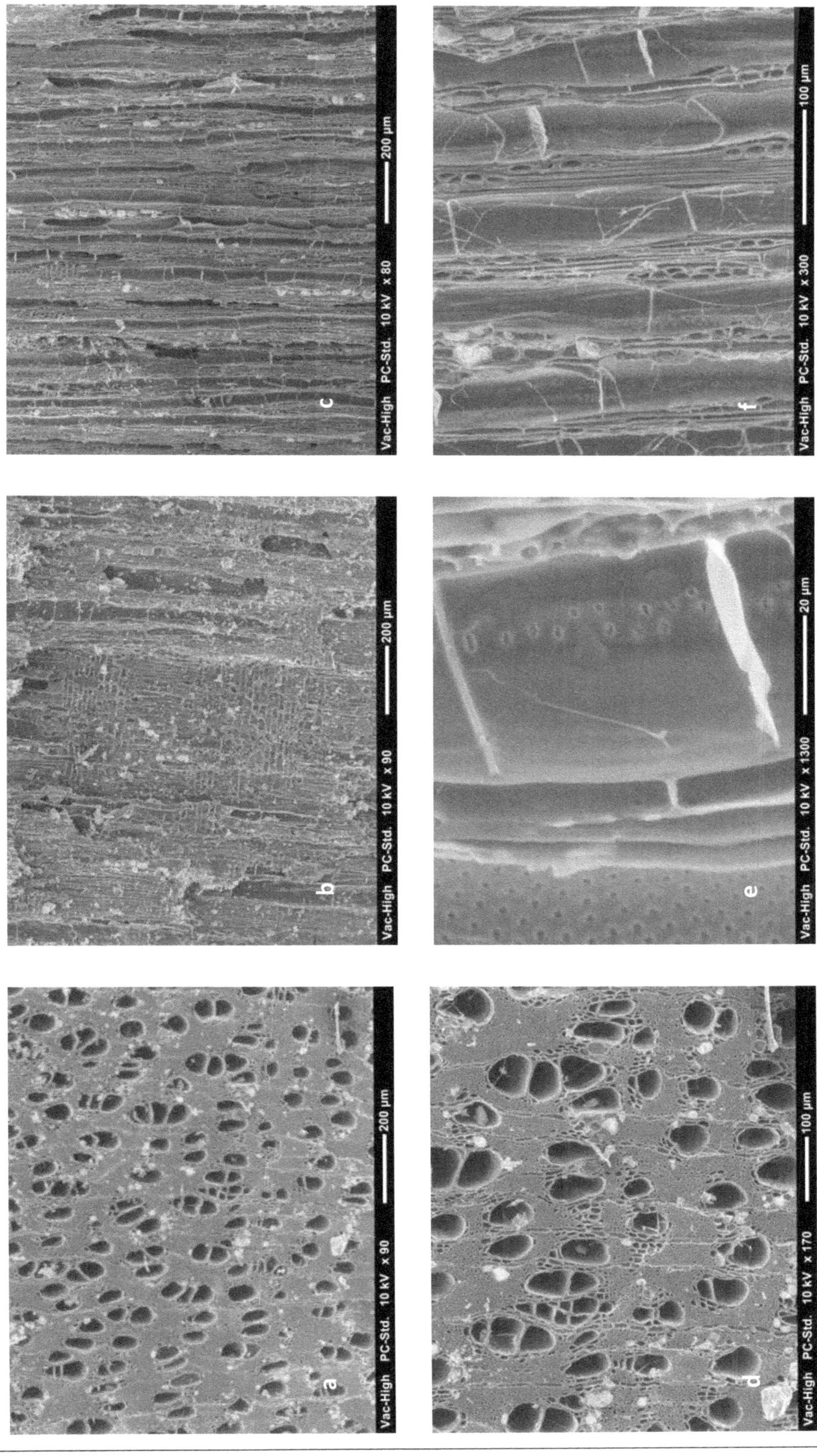

a and d: transversal, b and e: radial, c and f: tangential

A23 - OLACACEAE - *Ximenia americana* L. (Erongo, Namibia)

IAWA Feature Code:2,5,9,10, 13,22,24,41,47,53,56,61,65,79,83,97,102,104,115,136,142

- Growth ring boundaries absent
- Wood diffuse porous
- Vessels exclusively solitary (90% or more)
- Solitary vessel outlines circular
- Perforation plates simple
- Intervessel pits alternate
- Mean tangential diameter of vessel lumina: 50-100µm
- 20-40 vessels per square millimetre
- Fibres very thick-walled
- Axial parenchyma absent or extremely rare
- Four (3-4) cells per parenchyma strand
- Ray width 1-2 seriate
- Ray height: >1mm
- Radial canals present in rays
- Rays with procumbent, square and upright cells mixed throughout the ray
- Perforated fibre cells
- Rays: <=4-12 / mm
- Prismatic crystals present
- Druses present in fibre cells

Ximenia americana **Photo: A. Nankela**

a and d: transversal, b and e: radial, c and f: tangential

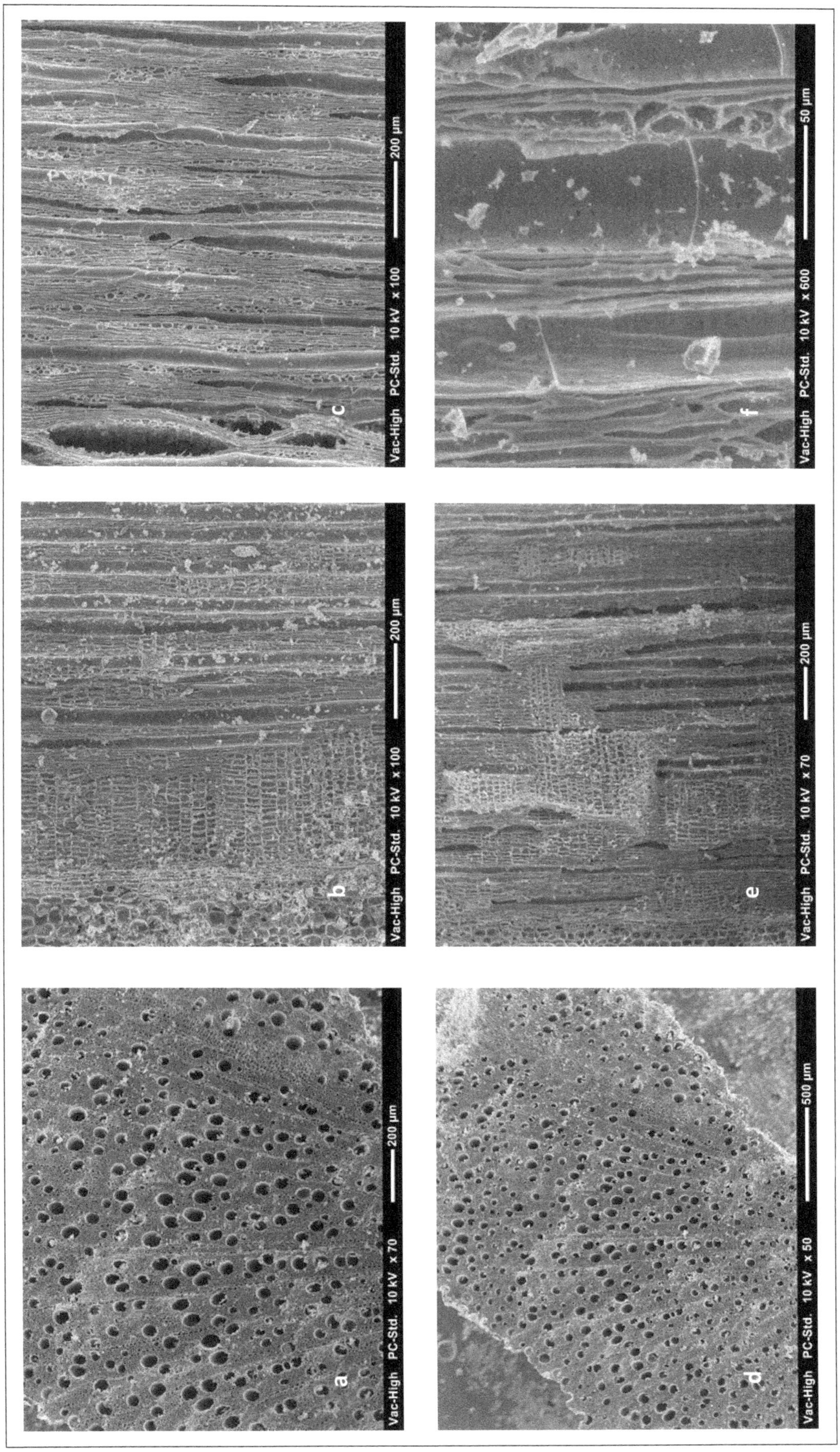

A24 – RHAMNACEAE - *Ziziphus mucronata* Willd. (Erongo, Namibia)

IAWA Feature Code: 2,5,9,22,24,41,48,49,70,65,92,96,101,105,114,147

- Growth ring boundaries absent
- Wood diffuse porous
- Vessels exclusively solitary (90% or more)
- Small radial multiples present
- Solitary vessel outlines angular(d)
- Perforation plates simple
- Intervessel pits alternate
- Intervessel pits: Minute - <= 4 µm
- Mean tangential diameter of vessel lumina: 50-100µm
- 20-40 vessels per square millimetre
- Fibres thin to thick-walled
- Axial parenchyma diffuse
- Four (3-4) cells per parenchyma strand
- Rays exclusively uniseriate
- Ray height: >1mm
- Rays with procumbent, square and upright cells mixed throughout the ray
- Perforated fibre cells
- Rays: <=4-12 / mm
- Septate fibres present
- Prismatic crystals present in ray cells
- Druses present in fibre cells
- Prismatic crystals in chambered upright and/or square ray cells

***Ziziphus mucronata*. Photo: M. Mvimi**

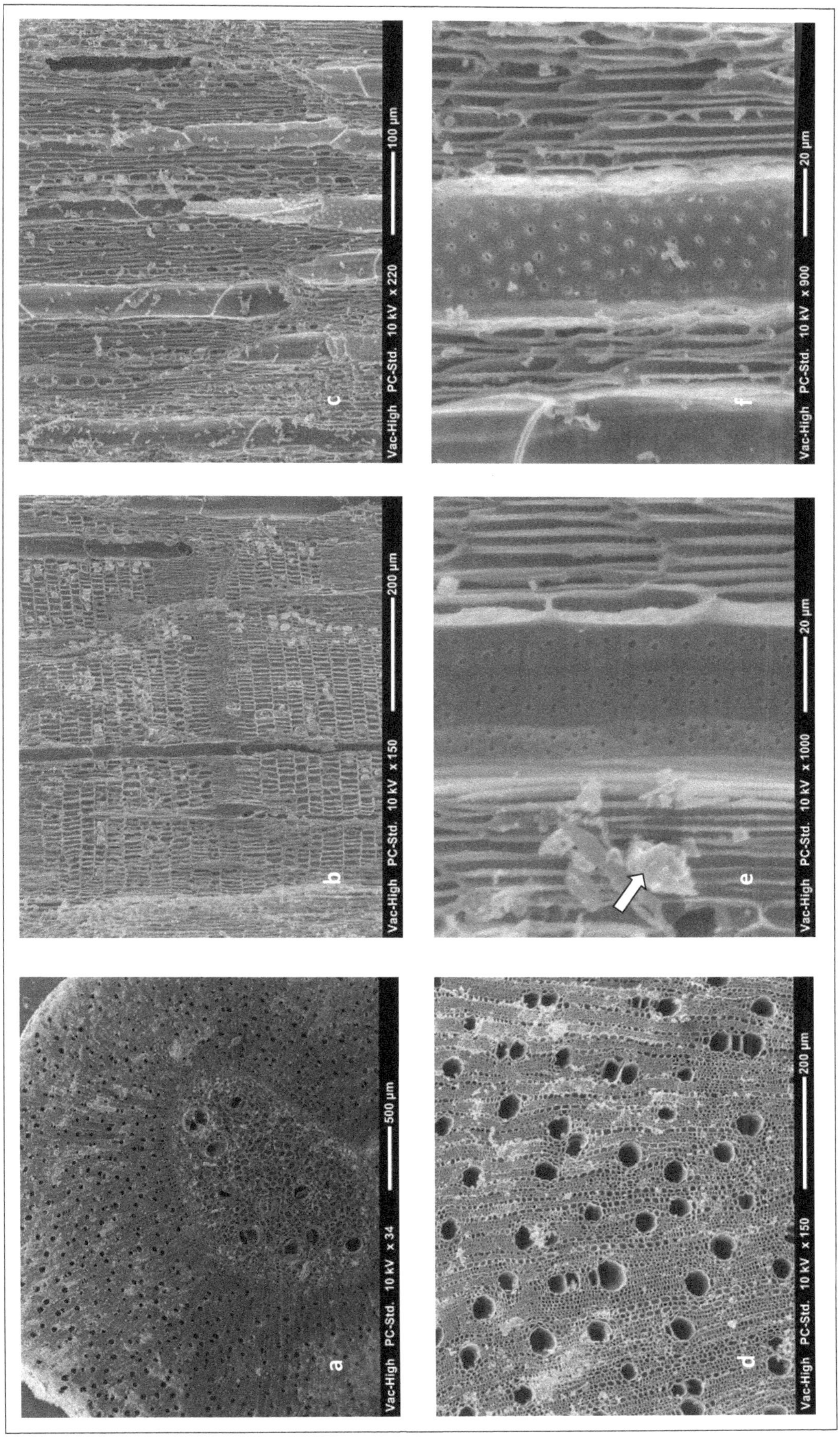
a
Vac-High PC-Std. 10 kV x 34
500 µm
b
Vac-High PC-Std. 10 kV x 150
200 µm
c
Vac-High PC-Std. 10 kV x 220
100 µm
d
Vac-High PC-Std. 10 kV x 150
200 µm
e
Vac-High PC-Std. 10 kV x 1000
20 µm
f
Vac-High PC-Std. 10 kV x 900
20 µm

A25 – RUBIACEAE - *Vangueria infausta Burch.* (Erongo, Namibia)

IAWA Feature Code: 2,5,9,13,22,41,49,65,68,75,97,102,109,115

- Growth ring boundaries absent
- Wood diffuse porous
- Vessels exclusively solitary (90% or more)
- Small pairs or radial multiples are present
- Solitary vessel outlines angular
- Perforation plates simple
- Intervessel pits alternate
- Minute - <= 4 µm
- Mean tangential diameter of vessel lumina: 50-100µm
- 40-100 vessels per square millimetre
- Fibres thin-walled
- Septate fibres present
- Axial parenchyma absent or extremely rare
- Four (3-4) cells per parenchyma strand
- Rays width 1-3 cells
- Rays of two distinct sizes
- Ray height: >1mm
- Rays with procumbent, square and upright cells mixed throughout the ray
- Rays: <=4-12 / mm

***Vangueria infausta*. Photo: M. Mvimi**

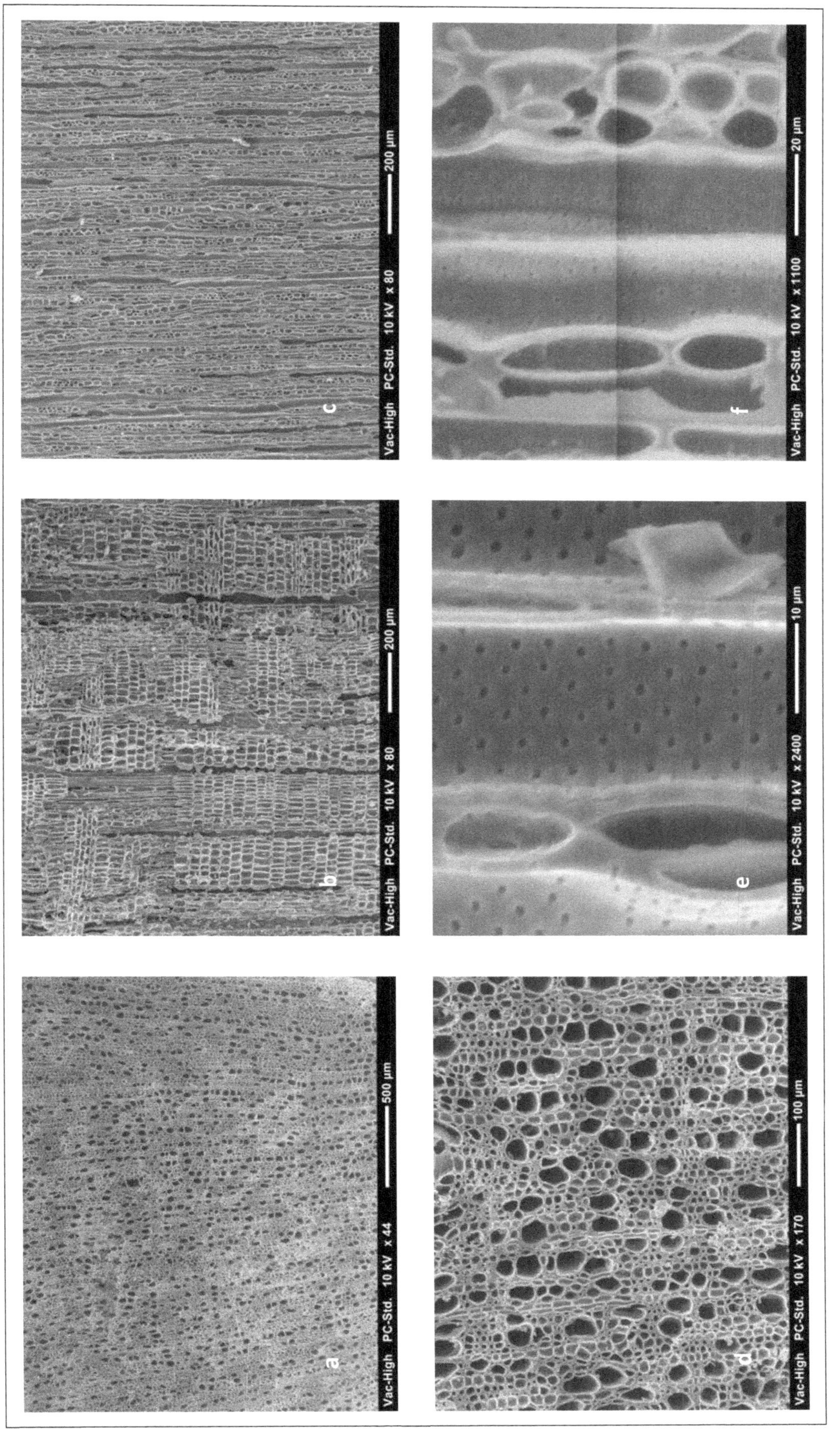

a and d: transversal, b and e: radial, c and f: tangential

A26 – FABACEAE - *Cassia abbreviata* Oliv. (Southeastern Botswana)

IAWA Feature Code: 2,5,9,13,22,41,46,65,69,85,93,97,109,116

- Growth ring boundaries absent
- Wood diffuse porous
- Vessels exclusively solitary (90% or more)
- Solitary vessel outlines circular
- Perforation plates simple
- Intervessel pits alternate
- Mean tangential diameter of vessel lumina: 50-100μm
- <= 5 vessels per square millimetre
- Fibres thin to thick-walled
- Axial parenchyma bands more than three cells wide
- Eight (5-8) cells per parenchyma strand
- Rays 1 to 2 cells
- Ray height: >1mm
- Rays with procumbent, square and upright cells mixed throughout the ray
- Perforated fibre cells
- Rays: <=4-12 / mm
- Septate fibres present

***Cassia abbreviata*. Photo: M. Mvimi**

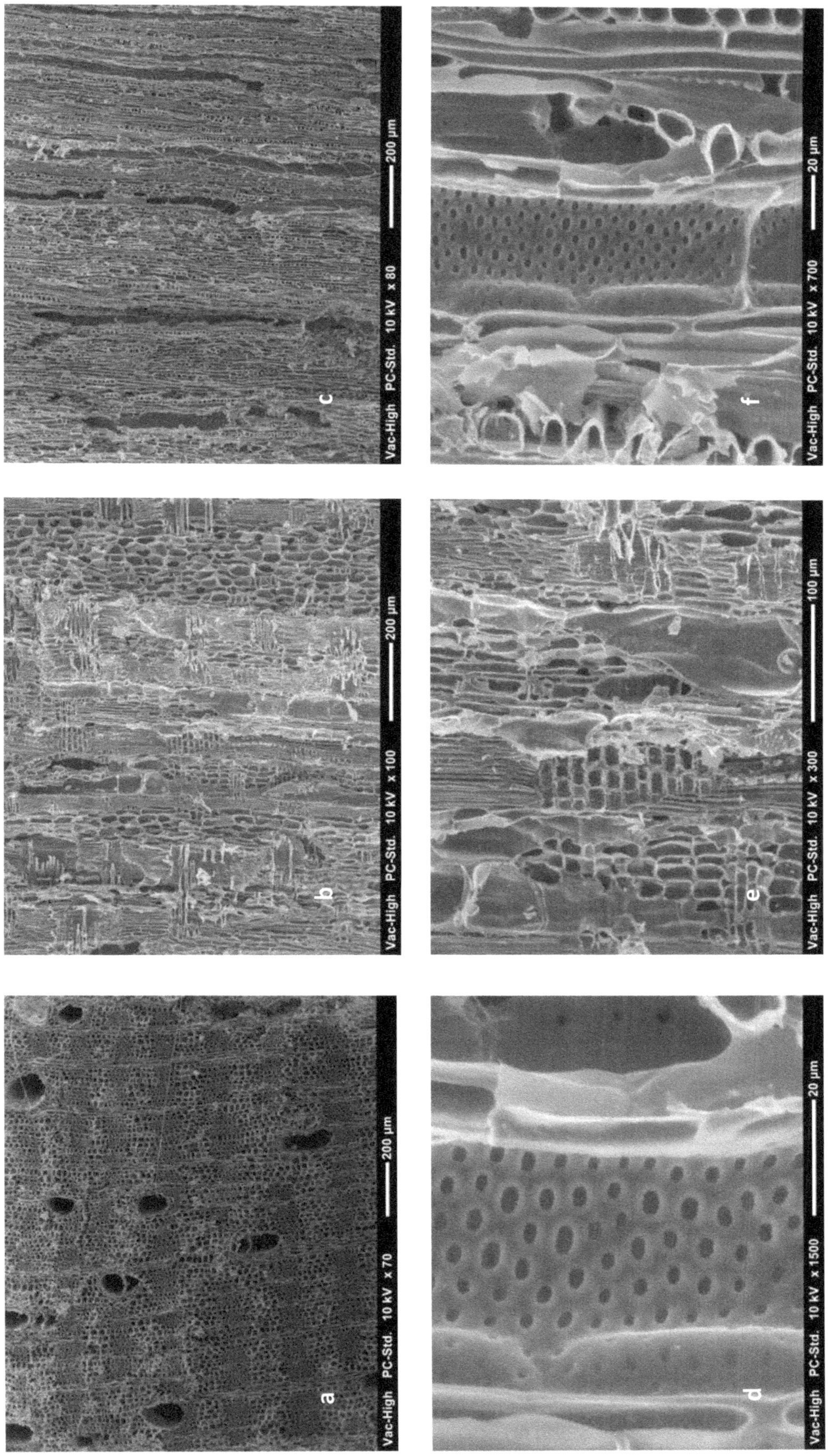

a: transversal, b and e: radial, c, d and f: tangential

References

Abagond.wordpress.com. (n.d.). *Bantu expansion.* Retrieved May 18, 2018, from https://abagond.wordpress.com/2018/02/27/bantu-expansion/

Angula, M. (2010). Gender and Climate Change: Namibia Case Study. *Heinrich Böll Stiftung Southern Africa, January*, 1–50. https://doi.org/10.1038/srep23539

Armstrong, W. . (n.d.). *Unforgettable Acacias A Large Genus Of Trees & Shrubs*. Retrieved May 18, 2018, from https://www2.palomar.edu

Badenhorst, S. (2006). Goats (Capra hircus), the Khoekhoen and Pastoralism: Current evidence from Southern Africa. *African Archaeological Review*, *23*(3–4), 45–53. https://doi.org/10.1007/s10437-006-9007-0

Barbieri, C., Vicente, M., Oliveira, S., Bostoen, K., Rocha, J., Stoneking, M., & Pakendorf, B. (2014). Migration and interaction in a contact zone: mtDNA variation among Bantu-speakers in Southern Africa. *PLoS ONE*, *9*(6). https://doi.org/10.1371/journal.pone.0099117

Barnard, A. (2008). Ethnographic analogy and the reconstruction of early Khoekhoe society. *Southern African Humanities*, *20*, 61–75.

Bationo, A., Hartemink, A., Lungu, O., Naimi, M., Okoth, P., Smaling, E., Thiombiano, L., & Waswa, B. (2012). *Improving Soil Fertility Recommendations in Africa using the Decision Support System for Agrotechnology Transfer (DSSAT)*. https://doi.org/10.1007/978-94-007-2960-5

Behar, D. M., Villems, R., Soodyall, H., Blue-Smith, J., Pereira, L., Metspalu, E., Scozzari, R., Makkan, H., Tzur, S., Comas, D., Bertranpetit, J., Quintana-Murci, L., Tyler-Smith, C., Wells, R. S., & Rosset, S. (2008). The Dawn of Human Matrilineal Diversity. *American Journal of Human Genetics*, *82*(5), 1130–1140. https://doi.org/10.1016/j.ajhg.2008.04.002

Bizimenyera, E. S., Githiori, J. B., Eloff, J. N., & Swan, G. E. (2006). In vitro activity of Peltophorum africanum Sond. (Fabaceae) extracts on the egg hatching and larval development of the parasitic nematode Trichostrongylus colubriformis. *Veterinary Parasitology*, *142*(3–4), 336–343. https://doi.org/10.1016/j.vetpar.2006.06.013

Bousman, C. B., & Brink, J. S. (2018). The emergence, spread, and termination of the Early Later Stone Age event in South Africa and southern Namibia. *Quaternary International*, *495*, 116–135. https://doi.org/10.1016/j.quaint.2017.11.033

Brenner, M. J., Ryano, K. P., & Wurz, S. (2020). Coastal adaptation at Klasies River main site during MIS 5c-d (93,000–110,000 years ago) from a southern Cape perspective. *Journal of Island and Coastal Archaeology*, *0*(0), 1–28. https://doi.org/10.1080/15564894.2020.1774444

Brown, T. A., Sekonya, K. G., Sadr, K., Sideras-Haddad, E., Bousman, C. B., & Smith, A. B. (2017). New radiocarbon dates and the herder occupation at Kasteelberg B, South Africa. *Antiquity*, *91*(359), 1299–1313. https://doi.org/10.15184/aqy.2017.102

Burrough, S. L., Thomas, D. S. G., & Singarayer, J. S. (2009). Late Quaternary hydrological dynamics in the Middle Kalahari: Forcing and feedbacks. *Earth-Science Reviews*, *96*(4), 313–326. https://doi.org/10.1016/j.earscirev.2009.07.001

Carrión, J. S., Scott, L., Huffman, T., & Dreyer, C. (2000). Pollen analysis of Iron Age cow dung in southern Africa. *Vegetation History and Archaeobotany*, *9*, 239–249. https://doi.org/10.1007/BF01294638

Chase, B. M., & Meadows, M. E. (2007). Late Quaternary dynamics of southern Africa's winter rainfall zone. *Earth-Science Reviews*, *84*(3–4), 103–138. https://doi.org/10.1016/j.earscirev.2007.06.002

Christenhusz, M. J. M., & Byng, J. W. (2016). Phytotaxa. *Phytotaxa*, *261*(3), 201–217. https://doi.org/10.11646/phytotaxa.261.3.1

Clark, J. D. (1959). *The prehistory of southeren Africa.* Penguin Books.

Clark, J. D., & Walton, J. (1962). A Late Stone Age Site in the Erongo Mountains South West Africa. *Proceedings of the Prehistoric Society*, *28*(I), 1–16. https://doi.org/10.1017/S0079497X00015632

Cordova, C. E., Scott, L., Chase, B. M., & Chevalier, M. (2017). Late Pleistocene-Holocene vegetation and climate change in the Middle Kalahari, Lake Ngami, Botswana. *Quaternary Science Reviews*, *171*, 199–215. https://doi.org/10.1016/j.quascirev.2017.06.036

d'Errico, F., & Backwell, L. (2016). Earliest evidence of personal ornaments associated with burial: The Conus shells from Border Cave. *Journal of Human Evolution*, *93*, 91–108. https://doi.org/10.1016/j.jhevol.2016.01.002

Dabney, C., Ostergaard, J., Watkins, E., & Chen, C. (2016). A novel method to characterize silica bodies in grasses. *Plant Methods*, *12*(1). https://doi.org/10.1186/s13007-016-0108-8

Daemane, M. E., Cilliers, S. S., & Bezuidenhout, H. (2010). An ecological study of the plant communities in the proposed Highveld National Park, in the peri-urban area of Potchefstroom, South Africa. *Koedoe*, *52*(1), 1–8. https://doi.org/10.4102/koedoe.v52i1.708

Dan, V., Mchombu, K., & Mosimane, A. (2010). Indigenous medicinal knowledge of the san people: The case of farm six, Northern Namibia. *Information Development, 26*(2), 129–140. https://doi.org/10.1177/0266666910367479

Deacon, H. J. (2017). *South African Archaeological Society A Review of the Post-Pleistocene in South Africa Author (s): H . J . Deacon Source : Goodwin Series , No . 1 , The Interpretation of Archaeological Evidence (Jun ., 1972), Published by : South African Archaeologica. 1*, 26–45.

Deacon, J. (1984). Later Stone Age people and their descendants in southern Africa. In: Klein, R. (ed.) *Southern African Prehistory and Paleoenvironment:* 221–328. Rotterdam: Balkema.

Demangeon, A. (2018). *Le Kalahari , d ' après le livre de Mr Siegfried Passarge*. 43–58.

Dewitte, O., Jones, A., Bosco, C., Spaargaren, O., & Montanarella, L. (2010). Soil Mapping for Africa : a Contemporary Perspective. *Bulletin de La Société Géographique de Liège, 54*(March 2015), 107–116.

Dotte-Sarout, E., Carah, X., & Byrne, C. (2015). Not just carbon: assessment and prospects for the application of anthracology in Oceania. *Archaeology in Oceania, 50*(1), 1–22. https://doi.org/10.1002/arco.504

Dunne, J., Mercuri, A. M., Evershed, R. P., Bruni, S., & Di Lernia, S. (2016). Earliest direct evidence of plant processing in prehistoric Saharan pottery. *Nature Plants, 3*(December). https://doi.org/10.1038/nplants.2016.194

Eichhorn, B., & Vogelsang, R. (2011). *Under the mopane tree: Holocene settlement in northern Namibia* (R. Kuper (Ed.). Heinrich Barth Institut.

Eitel, B., Eberle, J., & Kuhn, R. (2002). Holocene environmental change in the Otjiwarongo thornbush savanna (Northern Namibia): Evidence from soils and sediments. *Catena, 47*(1), 43–62. https://doi.org/10.1016/S0341-8162(01)00170-9

Eitel, B., Kadereit, A., Blümel, W. D., Hüser, K., Lomax, J., & Hilgers, A. (2006). Environmental changes at the eastern Namib Desert margin before and after the Last Glacial Maximum: New evidence from fluvial deposits in the upper Hoanib River catchment, northwestern Namibia. *Palaeogeography, Palaeoclimatology, Palaeoecology, 234*(2–4), 201–222. https://doi.org/10.1016/j.palaeo.2005.10.015

Elmoutaki, S. (1994). *Transition glaciaire-interglaciaire et Younger Dryas dans l'Hémisphère Sud (1°–20° Sud): analyse palynologique à haute résolution de sondages marin et continentaux (lac et marécage)*. University of Aix- Marseille III.

Esterhuysen, A., & Mitchell, P. (1996). Palaeoenvironmental and archaeological implications of charcoal assemblages from Holocene sites in western Lesotho, southern Africa. Palaeoecology of Africa, 24(May), 203–232.

Fitchett, J. M., Bamford, M. K., Mackay, A. W., & Grab, S. W. (2017). Chrysocoma ciliata L. (Asteraceae) in the Lesotho Highlands: an anthropogenically introduced invasive or a niche coloniser? *Biological Invasions, 19*(9), 2711–2728. https://doi.org/10.1007/s10530-017-1478-1

Forssman, T. (2019). A Review of Hunter–Gatherers in Later Stone Age Research in Southern Africa. *South African Archaeological Society Goodwin Series* 12: 56–68.

Gibbs Russell, G. E. (1987). Preliminary floristic analysis of the major biomes in southern Africa. *Bothalia, 17*(2), 213–227. https://doi.org/10.4102/abc.v17i2.1038

Gil-Romera, G., Scott, L., Marais, E., & Brook, G. A. (2007). Late Holocene environmental change in the northwestern Namib Desert margin: New fossil pollen evidence from hyrax middens. *Palaeogeography, Palaeoclimatology, Palaeoecology, 249*(1–2), 1–17. https://doi.org/10.1016/j.palaeo.2007.01.002

Goodwin, A. J. (1926). The hardy collection of stone implements. *South African Journal of Science,* XXIII, 826–832.

Goodwin, A. J. (1929). The Stone Ages in South Africa. *Africa, 2*(02), 174–182. https://doi.org/10.2307/1155825

Haacke, Wilfrid, H. G. (2008). Linguistic hypotheses on the origin of Namibian Khoekhoe speakers. *Southern African Humanities, 20*(2007), 163–177.

Heine, B., & König, C. (2008). What can linguistics tell us about early Khoekhoe history? *Southern African Humanities, 20*, 235–248.

Heine, K. (2005). Holocene Climate of Namibia: a Review Based on Geoarchives. *African Study Monographs, 30*(March), 119–133.

Henn, B. M., Gignoux, C., Lin, A. a, Oefner, P. J., Shen, P., Scozzari, R., Cruciani, F., Tishkoff, S. a, Mountain, J. L., & Underhill, P. a. (2008). Y-chromosomal evidence of a pastoralist migration through Tanzania to southern Africa. *Proceedings of the National Academy of Sciences of the United States of America, 105*(31), 10693–10698. https://doi.org/10.1073/pnas.0801184105

Huffman, T. (2007). *Handbook of the Iron Age*. University of KwaZulu-Natal Press.

Huffman, T. N. (1994). Toteng Pottery and the Origins of Bambata. *Southern African Field Archaeology, 3*, 3–9.

Humphries, M. S., Benitez-Nelson, C. R., Bizimis, M., & Finch, J. M. (2017). An aeolian sediment reconstruction of regional wind intensity and links to larger scale climate variability since the last deglaciation from the east coast of southern Africa. *Global and Planetary*

Change, *156*, 59–67. https://doi.org/10.1016/j.gloplacha.2017.08.002

Jacobson, L. (1987). The Size Variability of Ostrich Eggshell Beads from Central Namibia and Its Relevance as a Stylistic and Temporal Marker. *The South African Archaeological Bulletin*, *42*(145), 55–58.

Jacobson, L. (1987). More on Ostrich Eggshell Bead Size Variability: The Geduld Early Herder Assemblage. *South African Archaeological Bulletin*, *42*(146), 174–175.

Jolly, P. (1996). Symbiotic Interaction Between Black Farmers and South-Eastern San: Implications for Southern African Rock Art Studies, Ethnographic Analogy, and Hunter-Gatherer Cultural Identity. *Current Anthropology*, *37*(2), 277–305.

Kennedy, A. D., & Potgieter, A. L. F. (2003). Fire season affects size and architecture of Colophospermum mopane in southern African savannas. *Plant Ecology*, *167*(2), 179–192. https://doi.org/10.1023/A:1023964815201

Kinahan, J. (1995). Much Ado About Herding At Geduld : a Response To Smith and Jacobson. *South African Archaeological Bulletin*, *50*(162), 176–177.

Kurtz, D. V. (1994). Winnowing the "Great Kalahari Debate"; Its Impact on Hunter-Gatherer Studies: A Review of Current Literature. *PoLAR: Political Legal Anthropology Review* 17(1): 67–80. doi: 10.1525/pol.1994.17.1.67.

L Jacobson. (2010). *More on Ostrich Eggshell Bead Size Variability: The Geduld Early Herder Assemblage*. *42*(146), 174–175.

Le Meillour, L., Zazzo, A., Lesur, J., Cersoy, S., Marie, A., Lebon, M., Pleurdeau, D., & Zirah, S. (2018). Identification of degraded bone and tooth splinters from arid environments using palaeoproteomics. *Palaeogeography, Palaeoclimatology, Palaeoecology*, *511*, 472–482. https://doi.org/10.1016/j.palaeo.2018.09.013

Le Meillour, L., Zirah, S., Zazzo, A., Cersoy, S., Détroit, F., Imalwa, E., Lebon, M., Nankela, A., Tombret, O., Pleurdeau, D., & Lesur, J. (2020). Palaeoproteomics gives new insight into early southern African pastoralism. *Scientific Reports*, *10*(1), 1–11. https://doi.org/10.1038/s41598-020-71374-3

Lee, R.B (1984). *The Dobe !Kung*. Retrieved May 18, 2018, from https://laulima.hawaii.edu

Lim, S., Chase, B. M., Chevalier, M., & Reimer, P. J. (2016). 50,000 years of vegetation and climate change in the southern Namib Desert, Pella, South Africa. *Palaeogeography, Palaeoclimatology, Palaeoecology*. https://doi.org/10.1016/j.palaeo.2016.03.001

Ludemann, T., & Nelle, O. (2017). Anthracology: local to global significance of charcoal science. *Quaternary International*, *457*(November 2017), 1–5. https://doi.org/10.1016/j.quaint.2017.10.003

Mafongoya, P. L., Peerbhay, K., Jiri, O., & Nhamo, N. (2017). Climate Scenarios in Relation to Agricultural Patterns of Major Crops in Southern Africa. In *Smart Technologies for Sustainable Smallholder Agriculture: Upscaling in Developing Countries*. Elsevier Inc. https://doi.org/10.1016/B978-0-12-810521-4.00002-5

Mauran, G. (2020). *Matières colorantes ferrugineuses, pigments de l'art rupestre et comportements des populations Later Stone Age à Leopard Cave (Erongo, Namibie)* (Issue December).

Mauran, G., Lebon, M., Détroit, F., Caron, B., Nankela, A., Pleurdeau, D., & Bahain, J. J. (2019). First in situ pXRF analyses of rock paintings in Erongo, Namibia: results, current limits, and prospects. *Archaeological and Anthropological Sciences*, *11*(8), 4123–4145. https://doi.org/10.1007/s12520-019-00787-7

Mcbrearty, S., & Brooks, A. S. (2000). The revolution that wasn't: A new interpretation of the origin of modern human behavior. *Journal of Human Evolution*, *39*(5), 453–563. https://doi.org/10.1006/jhev.2000.0435

McFarlane, M. . J. ., & Eckardt, F. . D. . (2006). Lake Deception : A New Makgadikgadi Palaeolake. *Botswana Notes and Records*, *38*(2006), 195–201.

Mendelsohn, J., (2008). *Customary and legislative aspects of land registration and management on communal land in Namibia*, Ministry of Lands and Resettlement, Windhoek.

Metwally, A. A., Scott, L., Neumann, F. H., Bamford, M. K., & Oberhänsli, H. (2014). Holocene palynology and palaeoenvironments in the Savanna Biome at Tswaing Crater, central South Africa. *Palaeogeography, Palaeoclimatology, Palaeoecology*. https://doi.org/10.1016/j.palaeo.2014.03.019

Mitchell, P.J. (1997). Holocene Later Stone Age hunter-gatherers south of the Limpopo River, ca. 10,000–2,000 B.P. *Journal of World Prehistory*, 11, 359–424.

Mitchell, P. (2002a). Hunter-gatherer archaeology in southern Africa: recent research, future trends. Before Farming, 1(3), 1–36. https://doi.org/10.3828/bfarm.2002.1.3

Mitchell, P. (2002b). *The Archaeology of Southern Africa*. Cambridge University Press.

Mitchell, P. (2010). Genetics and southern African prehistory:an archaeological view. *Journal of Anthropological Sciences*, *88*, 73–92.

Mujaji, C. (n.d.). Diversity of landraces and wild forms of watermelon (Citrullus lanatus) in southern Africa A synopsis of the PhD Study Diversity of landraces and wild forms of watermelon (Citrullus lanatus) in southern Africa A synopsis of the PhD Study. In *Sciences-New York* (Issue November 2009).

Neumann, K. (1992). Bulletin de la Société Botanique de France. Actualités Botaniques The contribution of anthracology to the study of the late Quaternary vegetation history of the Mediterranean region and Mrica. *BuU. Soc. Bot. Fr. Actual. Bot*, *139*(February 2015), 421–440. https://doi.org/10.1080/01811789.1992.10827118

Nicholson, S. E. (2000). The nature of rainfall variability over Africa on time scales of decades to millenia. *Global and Planetary Change*, *26*(1–3), 137–158. https://doi.org/10.1016/S0921-8181(00)00040-0

Palmer, A. R., & Staden, J. M. (2006). Predicting the distribution of plant communities using annual rainfall and elevation: an example from southern Africa. *Journal of Vegetation Science*, *3*(2), 261–266. https://doi.org/10.2307/3235688

Pickering, R., Kramers, J. D., Hancox, P. J., de Ruiter, D. J., & Woodhead, J. D. (2011). Contemporary flowstone development links early hominin bearing cave deposits in South Africa. *Earth and Planetary Science Letters*, *306*(1–2), 23–32. https://doi.org/10.1016/j.epsl.2011.03.019

Pleurdeau, D. et al. (2018). *No Title: Rapport de la Mission Archéologique en Namibie.*

Pleurdeau, D., Imalwa, E., Détroit, F., Lesur, J., Veldman, A., Bahain, J. J., & Marais, E. (2012). "Of sheep and men": Earliest direct evidence of caprine domestication in southern Africa at Leopard Cave (Erongo, Namibia). *PLoS ONE*, *7*(7), 1–10. https://doi.org/10.1371/journal.pone.0040340

Plug, I. (1997). Late Pleistocene and Holocene hunter-gatherers in the Eastern Highlands of South Africa and Lesotho: A faunal interpretation. *Journal of Archaeological Science*, *24*(8), 715–727. https://doi.org/10.1006/jasc.1996.0153

Prior, J., & Williamsb, D. P. (1985). *An Investigation of Climatic Change in the Holocene Epoch using Archaeological Charcoal from Swaziland , Southern Africa.*

Prychid, C., Gardens, R. B., Rudall, P. J., & Gardens, R. B. (2004). *Systematics and Biology of Silica Bodies in Monocotyledons Systematics and Biology of Silica Bodies in Monocotyledons*. *8101*(October 2003). https://doi.org/10.1663/0006-8101(2004)069

Ramberg, L., Hancock, P., Lindholm, M., Meyer, T., Ringrose, S., Sliva, J., Van As, J., & VanderPost, C. (2006). Species diversity of the Okavango Delta, Botswana. *Aquatic Sciences*, *68*(3), 310–337. https://doi.org/10.1007/s00027-006-0857-y

Rapold, C., & Widlok, T. (2008). Dimensions of variability in Northern Khoekhoe language and culture. *Southern African Humanities*, *20*(1997), 133–161.

Raven, P., Evert, R., and Eichhorn, S. (2005). *Biology of Plants* (7th ed.). W.H. Freeman, New York.

Reynard, J. P., & Wurz, S. (2020). The palaeoecology of Klasies River, South Africa: An analysis of the large mammal remains from the 1984–1995 excavations of Cave 1 and 1A. *Quaternary Science Reviews*, *237*(July), 106301. https://doi.org/10.1016/j.quascirev.2020.106301

Robbins, L. H., Campbell, A. C., Murphy, M. L., Brook, G. A., Liang, F., Skaggs, S. A., Srivastava, P., Mabuse, A. A., & Badenhorst, S. (2008). Recent archaeological research at Toteng, Botswana: Early domesticated livestock in the Kalahari. *Journal of African Archaeology*, *6*(1), 131–149. https://doi.org/10.3213/1612-1651-10106

Robbins, L. H., Campbell, A. C., Murphy, M. L., Brook, G. a., Srivastava, P., & Badenhorst, S. (2005). The Advent of Herding in Southern Africa: Early AMS Dates on Domestic Livestock from the Kalahari Desert. *Current Anthropology*, *46*(4), 671–677. https://doi.org/10.1086/432748

Robbins, L., Murphy, M., Campbell, A., Brook, G., Reid, D., Haberyan, A., & Downey, W. (1998). Test excavations and reconnaissance palaeoenvironmental work at Toteng, Botswana. *The South African Archaeological Bulletin*, *53*(168), 125–132.

Rutherford, M. C., Mucina, L., & Powrie, L. W. (2006). Biomes and Bioregions of Southern Africa. *The Vegetation of South Africa, Lesotho and Swaziland. Strelitzia 19*, 30–51. https://doi.org/10.1289/ehp.7863

Sadr, K. (1998). The First Herders at the Cape of Good Hope. *African Archaeological Review*, *15*(2), 101–132.

Sadr, K. (2008). Invisible herders? The archaeology of Khoekhoe pastoralists. *South African Humanities*, *20*, 179–203.

Sadr, K., Smith, A., Plug, I. N. A., Orton, J., Mütti, B., & Mutti, B. (2003). Herders and Foragers on Kasteelberg: Interim Report of Excavations 1999-2002. *The South African Archaeological Bulletin*, *58*(177), 27–32. https://doi.org/10.2307/3889154

Sadr K., (2003) The Neolithic of southern Africa. *Journal of African History*. 44: 195–209. Doi: 10.1017/S0021853702008393

Sarantou, M. A. (2018). *Namibian narratives : postcolonial identities in craft and design (Magister Technologiae : Fashion) School for Art , Architecture and Design , University of South Australia 7 October 2014 Doctor of Philosophy (PhD): Visual Arts. October 2014.*

Schuster, S. C., Miller, W., Ratan, A., Tomsho, L. P., Giardine, B., Kasson, L. R., Harris, R. S., Petersen, D. C., Zhao, F., Qi, J., Alkan, C., Kidd, J. M., Sun, Y., Drautz, D. I., Bouffard, P., Muzny, D. M., Reid, J. G., Nazareth, L. V, Wang, Q., … Hayes, V. M. (2010). Complete Khoisan and Bantu genomes from southern Africa. *Nature*, *463*, 943. https://doi.org/10.1038/nature08795

Scott, L., Marais, E., & Brook, G. A. (2004). Fossil hyrax dung and evidence of Late Pleistocene and Holocene vegetation types in the Namib Desert. *Journal of*

Quaternary Science, 19(8), 829–832. https://doi.org/10.1002/jqs.870

Scott, L., & Neumann, F. H. (2018). Pollen-interpreted palaeoenvironments associated with the Middle and Late Pleistocene peopling of Southern Africa. *Quaternary International*. https://doi.org/10.1016/j.quaint.2018.02.036

Scott, L., & Woodborne, S. (2007a). Vegetation history inferred from pollen in Late Quaternary faecal deposits (hyraceum) in the Cape winter-rain region and its bearing on past climates in South Africa. *Quaternary Science Reviews*. https://doi.org/10.1016/j.quascirev.2006.12.012

Scott, L., & Woodborne, S. (2007b). Vegetation history inferred from pollen in Late Quaternary faecal deposits (hyraceum) in the Cape winter-rain region and its bearing on past climates in South Africa. *Quaternary Science Reviews, 26*(7–8), 941–953. https://doi.org/10.1016/j.quascirev.2006.12.012

Shaw, P. (1985). Late Quaternary Landforms and Environmental Change in Northwest Botswana: The Evidence of Lake Ngami the Mababe Depression. *Transactions of the Institute of British Geographers, 10*(3), 333–346. https://doi.org/10.2307/622182

Shaw, P. A., Bateman, M. D., Thomas, D. S. G., & Davies, F. (2003). Holocene fluctuations of Lake Ngami, Middle Kalahari: Chronology and responses to climatic change. *Quaternary International, 111*(1), 23–35. https://doi.org/10.1016/S1040-6182(03)00012-0

Shillito, L., & Almond, M. J. (2010). *Comment on : Fruit and seed biomineralization and its effect on preservation by E . Messager et al .; in : Archaeological*. 225–229. https://doi.org/10.1007/s12520-010-0041-0

Shi, N., Dupont, L. M., Beug, H., & Schneider, R. (1998). Vegetation and climate changes during the last 21 000 years in S . W . Africa based on a marine pollen record. *Vegetation History and Archaeobotany* , 7, 127–140.

Sievers, C. (2006). Seeds from the Middle Stone Age layers at Sibudu Cave. *Southern African Humanities, 18*(1), 203–222.

Sievers, C. (2015). Nuts for dinner Cladium mariscus in the Middle Stone Age at Sibudu, South Africa. *Transactions of the Royal Society of South Africa, 70*(3), 213–218. https://doi.org/10.1080/0035919X.2014.996919

Sievers, C. (2017). What ' s the rush ? Scanning electron micrographs of Juncus (Juncaceae) seeds. *Southern African Humanities, 25, 207–216*

Simonin, K. A., & Roddy, A. B. (2017). Genome downsizing, physiological novelty, and the global dominance of flowering plants. *BioRxiv*, 1–15. https://doi.org/10.1101/174615

Sletten, H. R., Railsback, L. B., Liang, F., Brook, G. A., Marais, E., Hardt, B. F., Cheng, H., & Edwards, R. L. (2013). A petrographic and geochemical record of climate change over the last 4600years from a northern Namibia stalagmite, with evidence of abruptly wetter climate at the beginning of southern Africa's Iron Age. *Palaeogeography, Palaeoclimatology, Palaeoecology, 376*, 149–162. https://doi.org/10.1016/j.palaeo.2013.02.030

Smit, G. N., & Rethman, N. F. G. (1998). Root biomass, depth distribution and relations with leaf biomass of Colophospermum mopane. *South African Journal of Botany, 64*(1), 38–43. https://doi.org/10.1016/S0254-6299(15)30825-5

Smith, A. B. (2008). Early herders in southern Africa: a synthesis. *Animals and People: Archaeozoological Papaers in Honour of Ina Plug*, 94–103.

Smith, A. B. (2013). *Notes and Comment Khoesaan Orthography *. 53*(167), 37–38.

Smith, A. B., Yates, R., Jacobson, L., Smith, A. B., Yates, R., & Jacobson, L. (2010). *GEDULD CONTRA KINAHAN *. 51*(163), 36–39.

Smith, A. B., Yates, R., Miller, D., Jacobson, L., & Evans, G. (1995). Excavations at Geduld and the Appearance of Early Domestic Stock in Namibia. *South African Archaeological Bulletin, 50*(161), 3–20. https://doi.org/10.2307/3889270

Smith, B. W., & Ouzman, S. (2004). Taking Stock: Identifying Khoekhoen Herder Rock Art in Southern Africa. *Current Anthropology, 45*(4), 499–526. https://doi.org/10.1086/422081

Solway, J. S., & Lee, R. B. (1990). Foragers, genuine or spurious? Situating the Kalahari San in history. *Current Anthropology, 31*(2), 109–146. https://doi.org/10.2307/2743583

Soodyall, H., & Jenkins, T. (1993). Mitochondrial DNA polymorphisms in negroid populations from Namibia: New light on the origins of the dama, herero and ambo. *Annals of Human Biology, 20*(5), 477–485. https://doi.org/10.1080/03014469300002872

Stynder, D. D. (2009). Craniometric evidence for South African Later Stone Age herders and hunter-gatherers being a single biological population. *Journal of Archaeological Science*. https://doi.org/10.1016/j.jas.2008.11.001

Stynder, D. D., Ackermann, R. R., & Sealy, J. C. (2007). Craniofacial variation and population continuity during the South African holocene. *American Journal of Physical Anthropology*. https://doi.org/10.1002/ajpa.20696

Su, P. (2012). Sorption of metal ions to wood, pulp, and bark materials. Doctoral thesis, Åbo Akademi Univeristy. ISBN 9789521227776

Sullivan, S. (1998). People, plants and practice in drylands: socio-political and ecological dimensions of resource-use by Damara farmers in north-west Namibia. PhD thesis, University College London.

Sullivan, S. (1999). Folk and formal, local and national: Damara knowledge and community conservation in southern Kunene, Namibia. *Cimbebasia, 15*(January 1999), 1–28.

Symes, C. T. (2012). Seed dispersal and seed banks in Aloe marlothii (Asphodelaceae). *South African Journal of Botany, 78*, 276–280. https://doi.org/10.1016/j.sajb.2011.06.008

Symes, C. T. (2008). Avian nectarivory and pollination in Aloe marlothii Berger: interactions between bird communities and a winter-flowering succulent. PhD Thesis, University of Pretoria, Pretoria.

Taylor, A. R. D., Howard, G. W., & Begg, G. W. (1995). Developing wetland inventories in southern Africa: A review. *Vegetatio, 118*(1–2), 57–79. https://doi.org/10.1007/BF00045191

Thackeray, J. F. (1979). An Analysis of Faunal Remains from Archaeological Sites in Southern South West Africa (Namibia). *South African Archaeological Society, 34*(129), 18–33.

Thomas, D. S. G., Brook, G., Shaw, P., Bateman, M., Haberyan, K., Appleton, C., Nash, D., McLaren, S., & Davies, F. (2003). Late Pleistocene wetting and drying in the NW Kalahari: An integrated study from the Tsodilo Hills, Botswana. *Quaternary International, 104*, 53–67. https://doi.org/10.1016/S1040-6182(02)00135-0

Thomas, D. S. G., & Shaw, P. A. (2002). Late Quaternary environmental change in central southern Africa: New data, synthesis, issues and prospects. *Quaternary Science Reviews, 21*(7), 783–797. Doi:10.1016/S0277-3791(01)00127-5

Thornton-Barnett, S. R. (2013). *Ancestral Pharmacopeias: A Paleoethnobotanical Assessment of Plant Use in the Western Free State, South Africa*. Master thesis. San Marcos, Texas: Texas State University.

Tooth, S., McCarthy, T. S., Brandt, D., Hancox, P. J., & Morris, R. (2002). Geological controls on the formation of alluvial meanders and floodplain wetlands: The example of the Klip River, eastern Free State, South Africa. *Earth Surface Processes and Landforms, 27*(8), 797–815. Doi: 10.1002/esp.353

Tooth, Stephen, & McCarthy, T. S. (2007). Wetlands in drylands: Geomorphological and sedimentological characteristics, with emphasis on examples from southern Africa. *Progress in Physical Geography, 31*(1), 3–41. https://doi.org/10.1177/0309133307073879

Truc, L., Chevalier, M., Favier, C., Cheddadi, R., Meadows, M. E., Scott, L., Carr, A. S., Smith, G. F., & Chase, B. M. (2013). Quantification of climate change for the last 20,000years from Wonderkrater, South Africa: Implications for the long-term dynamics of the Intertropical Convergence Zone. *Palaeogeography, Palaeoclimatology, Palaeoecology, 386*, 575–587. Doi: 10.1016/j.palaeo.2013.06.024

Urso, V., Signorini, M. A., & Bruschi, P. (2013). Survey of the ethnobotanical uses of Ximenia americana L. (mumpeke) among rural communities in South Angola. *Journal of Medicinal Plants Research, 7*(1), 7–18. Doi: 10.5897/JMPR11.492

Van Rooyen, N., & van Rooyen, M. W. (1998). Vegetation of the south-western arid kalahari: An overview. *Transactions of the Royal Society of South Africa, 53*(2), 113–140. https://doi.org/10.1080/00359199809520381

van Wyk, B., & van Wyk, P. (2007). *How To Identify Trees in Southern Africa by Van Wyk, Braam | Penguin Random House South Africa* (Gill Gordon (Ed.). Struik Publishers).

Voarintsoa, N. R. G., Brook, G. A., Liang, F., Marais, E., Hardt, B., Cheng, H., Edwards, R. L., & Railsback, L. B. (2017). Stalagmite multi-proxy evidence of wet and dry intervals in northeastern Namibia: Linkage to latitudinal shifts of the Inter-Tropical Convergence Zone and changing solar activity from AD 1400 to 1950. *Holocene*. Doi: 10.1177/0959683616660170

Vogelsang, R., Richter, J., Jacobs, Z., Eichhorn, B., Linseele, V., & Roberts, R. G. (2010). New excavations of Middle Stone Age deposits at Apollo 11 rockshelter, Namibia: Stratigraphy, archaeology, chronology and past environments. *Journal of African Archaeology, 8*(2), 185–218. Doi: 10.3213/1612-1651-10170

Wadley, L. (1979) Big Elephant Shelter: its role in the Holocene prehistory in central South Western Africa, *Cimbebasia (B)*. 3: 1-76.

Wadley, L. (2004). Vegetation changes between 61 500 and 26 000 years ago: The evidence from seeds in Sibudu Cave, KwaZulu-Natal. *South African Journal of Science, 100*(3–4), 167–173.

Wadley, L. (2006). Partners in grime: results of multi-disciplinary archaeology at Sibudu Cave. *Southern African Humanities, 18*(1), 315–341.

Wadley, L. (1976). Radiocarbon dates from Big Elephant Shelter , Erongo Mountains , South West Africa. *South African Archaeological Society*, 31, 146. Doi: 10.2307/3887735

Wadley, L., Sievers, C., Bamford, M., Goldberg, P., Berna, F., & Miller, C. (2011). Middle Stone Age bedding construction and settlement patterns at Sibudu, South Africa. *Science, 334*, 1388–1391. Doi: 10.1126/science.1213317

Walker, B. H. (1995). Conserving Biological Diversity through Ecosystem Resilience Walker, 1995.pdf. In *Conservation Biology*, 9(4), 747–752.

WENDT, W. E. (1972) Preliminary report on an archaeological research programme in South West Africa. *Cimbebasia (B) 2*, 1-61.

Wiedenhoeft, A. C., & Miller, R. B. (2005). Structure and Function of Wood. *Handbook of Wood Chemistry and Wood Composites*, 9–33. Doi:10.1201/b12487-4

Wikipedia. (2005). *Khoekhoe language*. Retrieved August 18, 2018, from https://en.wikipedia.org/wiki/Khoekhoe_language

Wilmsen, E. N., Denbow, J. R., Bicchieri, M. G., Binford, L. R., Gordon, R., Guenther, M., Lee, R. B., Ross, R., Solway, J. S., Tanaka, J., Vansina, J., & Yellen, J. E. (1990). Paradigmatic History of San-Speaking Peoples and Current Attempts at Revision [and Comments and Replies]. *Current Anthropology*, *31*(5), 489–524.

Wündsch, M., Haberzettl, T., Kirsten, K. L., Kasper, T., Zabel, M., Dietze, E., Baade, J., Daut, G., Meschner, S., Meadows, M. E., & Mäusbacher, R. (2016). Sea level and climate change at the southern Cape coast, South Africa, during the past 4.2 kyr. *Palaeogeography, Palaeoclimatology, Palaeoecology*. Doi: 10.1016/j.palaeo.2016.01.027

Yates, R., Miller, D., Jacobson, L., & Evans, G. (1995). Excavations at Geduld and the Appearance of Early Domestic Stock in Namibia. *South African Archaeological Bulletin*, *50*(161), 3–20. Doi: 10.2307/3889270

www.ingramcontent.com/pod-product-compliance
Ingram Content Group UK Ltd.
Pitfield, Milton Keynes, MK11 3LW, UK
UKHW061848041026
15591UKWH00015B/969